Climate Change

FOCUS SERIES

Series Editor Jean-Charles Pomerol

Climate Change

Identification and Projections

Philippe de Larminat

iSTE

WILEY

First published 2014 in Great Britain and the United States by ISTE Ltd and John Wiley & Sons, Inc.

ISTE Ltd
27-37 St George's Road
London SW19 4EU
UK

www.iste.co.uk

John Wiley & Sons, Inc.
111 River Street
Hoboken, NJ 07030
USA

www.wiley.com

Library of Congress Control Number: 2014950500

British Library Cataloguing-in-Publication Data
A CIP record for this book is available from the British Library
ISSN 2051-2481 (Print)
ISSN 2051-249X (Online)
ISBN 978-1-84821-777-5

Cover photo: Anna Rigoigne, 2013.

Contents

1

Introduction

1.1. Context

The IPCC (Intergovernmental Panel on Climate Change) was created in 1988 under the auspices of the UN. Its aim is the scientific study of the causes of the global warming observed over the course of the 20th Century, how it is likely to evolve in the future, its human and environmental consequences, and, subsequently, to give rise to appropriate policy decisions[1].

At the end of September 2013, at a plenary meeting in Stockholm, the IPCC presented a draft version of its fifth *assessment report from working group 1*: "Climate Change 2013, The Physical Science Basis". The *summary for policymakers*, as it is known, was debated and approved, in turn anticipating the approval of the report as a whole (October 2014, Copenhagen). These two documents, AR5 (Fifth Assessment Report) and SPM (Summary for Policy Makers)[2] embody the current expression of

1 The IPCC is not technically an international research body, like CERN. It is organized around a secretariat. Its experts are scientists from around the world who analyze and evaluate all of the research published in the field. Apart from issuing a report every five years, it coordinates a number of different operations, such as the Coupled Model Intercomparison Project (*CMIP5*: *Phase 5*).

2 The latter, of around 33 pages, is more well-known as it is more concise, and is approved in its entirety by government representatives and the scientists who wrote the report.

consensus in the scientific community. They are available on the IPCC website and are regularly referred to throughout this work[3].

According to the final press release, taken from the SPM (p. 17): "It is *extremely likely* that human influence has been the dominant cause of the warming observed since the mid-20th Century". Per the fourth report (AR4, 2007) this statement was only qualified as "*very likely*". In the highly standardized language of the IPCC, this means that its confidence in attributing such warming to human influence has increased from 90% to 95%.

This confidence is less evident in the texts. Of all the quantified evaluations in the SPM, without a doubt, the most significant is what is known as the planet's climate sensitivity. It quantifies the equilibrium temperature change that would be caused by a possible doubling of the concentration of CO_2 in the atmosphere[4]. According to the SPM (p. 14): "equilibrium climate sensitivity is *likely* in the range of 1.5°C to 4.5°C (*high confidence*), *extremely unlikely* less than 1°C (*high confidence)* and *very unlikely* greater than 6°C (*medium confidence*)."

According to the IPCC's future scenarios regarding concentration, the rate of CO_2 may well increase four-fold in the course of the next century. In the extreme cases, (1°C to 6°C), the respective consequences range from the minor to the catastrophic: 2°C or 12°C. Moreover, the *likely* range has broadened since 2007. The IPCC highlights the fact that "the lower limit of the likely range evaluated (1.5°C) is therefore lower than the 2°C stated in the AR4". Aware that the alarmist nature of the message may be diluted, scientists justified this modification to governmental delegates (p. 14): "this assessment reflects improved understanding of climate sensitivity, the extended temperature record in the atmosphere and the ocean, and new estimates of radiative forcing". Therefore, uncertainty has increased as knowledge has broadened, despite the 95% confidence stated. It is on this basis that international agreements are entered into, which involve the annual

3 http://www.climatechange2013.org/images/report/WG1AR5_SPM_FINAL.pdf https://www.ipcc.ch/report/ar5/wg1/

4 This simple formulation contains a logarithmic law: if a doubling of CO_2 leads to a temperature rise of S_{clim} (°C), a further doubling will not lead to 4 S_{clim}, but only 2 S_{clim}. A third doubling, eight times the initial level, would lead to a rise of 3 S_{clim}, and so on. Note that this is an empirical law where the extrapolation for high concentrations is questionable.

expenditure of several thousands of billions of dollars (a number of global GDP points).

Scientifically, the likelihoods mentioned above must not be taken at face value. Their levels are debated over so as to reach a consensus[5] with political figures (more than 190 governmental delegations were represented in Stockholm). Moreover, the IPCC states (AR5, 1.4.4) that they do not necessarily come from actual statistical calculations, but simply express the confidence experts have in their own judgments.

With this in mind, anything which can help to give a more exact evaluation of the planet's climatic parameters would be greatly appreciated. This is the case for model identification techniques, which is this author's field of expertise.

1.2. Identification

Identifying a process consists of determining a mathematical model, often reduced to external behavior, using the observation of input and output data (causes and effects). In the case of the climatic process, the relevant input data are: the atmospheric concentration of CO_2, the solar activity and the volcanic activity. The output is the overall surface temperature of the Earth. The theory on the identification of dynamic systems has been highly developed for several decades [LJU 87, LJU 99]. Seemingly all ingredients are available to be able to apply it to the overall climate system of the Earth: simple usable models, with a limited number of parameters, observations of input and output signals, and proven software toolboxes (Matlab®: System Identification toolbox).

One would therefore expect to find reams of studies on the subject. Yet this is not the case. The term *identification* (in the systemic sense) does not appear once in the 1,550 pages of the AR5, nor in the title of any of the 9,200 publications surveyed. On the internet, a key word search (identification, climate, model, etc.) does not provide any links. The only publication on the subject, that we are aware of, is entitled: "A fractal

5 Consensus (Larousse dictionary): "A procedure which gives rise to an agreement without the use of a formal vote, which avoids objections and abstentions". A definition or an oxymoron?

climate response function can simulate global average temperature trends of the modern era and the past millennium" by Van Hateren [VAN 13]. However, nothing in this title refers to identification. The key word which gets the closest is "modeling", and none of the bibliographic references given refer to the great Masters of identification theory (Aström, Ljung, Soderström, etc.). It is quite possible that the author is unknowingly applying identification, just as Jourdain used prose. With the exception of the excellent paper mentioned above, we couldn't find any other significant work on the global modeling of climatic process through identification.

Indeed, the IPCC has long been checking its models against the historical climate data available: its large numerical models, based on the laws of physics, as well as its simplified models, based on energy balances. Yet identification is not involved. At most, these models involve partial adjustments (closure parameters tuning) or fingerprinting (detection and attribution of anthropogenic impact).

According to Hervé le Treut (2004), Director of the *Institut Pierre Simon Laplace*[6]: "numerical models (i.e. large-scale physical models, simulated by digital calculators) play a key role in studies of the greenhouse effect because they are the *only tool* which can be used to evaluate future climates: the analogy with climates of past eras which experienced different CO_2 levels and the extrapolation toward the future of climatic data collected during the 20th Century provide unarguably precious information, but can only be interpreted with the help of physical models".

In this work, we look instead to push forward with the logic of identification, allowing the climatic data to speak for itself, using it as "black box" input and output data (causes and effects), without constraining it to any type of prior knowledge. This is not without its difficulties: the Earth's climatic process is at the limit of what can be identified. To achieve this goal, identification requires that input data be sufficiently accurate, with a suitable number of significant events. In this case, the effects caused by the input are partly obscured by the random fluctuations of the climate. Regarding CO_2, the first significant changes go back to less than a century ago, and their effects are difficult to distinguish from natural variations, both having the same order of magnitude, in terms of size and duration. Furthermore, to observe relatively large-scale temperature variations, it is

6 Participant institute in the CMIP 5.

necessary to look over more than a millennium, where uncertainties regarding paleoclimatic reconstructions are added to natural fluctuations. Moreover, the structure of the model must be finely-tuned to the objectives as well as to the identification method, otherwise the data will remain unreadable and analysis of uncertainties will remain difficult.

Nevertheless, this text shows that it is possible to obtain significant results in this way. It is therefore surprising that the community of climatologists is ignoring a technique which is taught in the first cycle of university courses, despite the fact that all the ingredients and application tools are readily available. It is also possible that in trialing such an approach, incoherent results were obtained, and therefore not published, or that results were self-censored as they were poorly-aligned with the other mainstream results presented by the IPCC.

1.3. Expectations and results

Above we criticized the fact that the current situation regarding physical knowledge does not allow us to accurately assess the planet's fundamental climatic parameters. Although unable to work miracles, identification can nevertheless provide results which can call into question the current scientific consensus on what is commonly referred to as "climate change".

Firstly, it will be argued that the assertion that the warming seen over the previous century is caused by human action is not confirmed, nor is it contradicted by observations. It therefore remains based solely on physical considerations, with a number of uncertainties to be addressed later (section 5.8). At the very least, identification can help to eliminate the extreme high values of climate sensitivity which have been forecast. This result falls short of expectations, but it would serve to counter the IPCC's familiar argument that the simple observation of climatic data gives evidence of the human influence on global warming.

Subsequently, the estimate of the sensitivity coefficient for solar activity and its range of uncertainty clearly show that fluctuations in solar activity constitute the predominant cause of recent global warming. The IPCC is opposed to this hypothesis, arguing that the variations in solar irradiance are too weak, and denying that there is any other way which the sun may have an effect. However, statistical analysis is clear: the sun can explain both large and small climatic variations, which can be observed despite the natural variability of the climate. This analysis relies on climatic databases which are, as a whole, accepted by the IPCC (AR5, Chapter 5).

Beyond statistical analysis, the predictive power of the models identified helps to confirm their validity. Solely on the basis of informations known in 2000, our models were able to provide a remarkably accurate reproduction of the "climatic pause" which occurred shortly after and which is continuing even now. It is not so with IPCC models: global temperatures observed systematically fall below the lower end of the range of short-term projections produced by these models, even when updated in 2006.

Regarding long-term predictions, they are highly dependent on the future of solar activity, and the author does not have the expertise necessary to assess the projections made by solar physics specialists. He is also unable to confirm or contradict the hypotheses on the forms of greenhouse effect, and the climate sensitivity coefficient which results from them. With the models identified, there is a wide range of extrapolations from the climate of the past millennium whereby, in the worst case scenario, a warming of two degrees compared to pre-industrial temperatures is unlikely to be reached by the end of the 21st Century, and the current climatic pause may be but an indication of a return to the little ice age of the 17th and 18th Centuries.

1.4. Contents of the work

The goal of this work is to describe our methodological approach with enough accuracy so that the reader, equipped with some knowledge on the theory of systems, in modeling and simulation[7], can initially verify its validity, and if necessary, reproduce and use them. Only Chapters 4 and 6 pose any problems to such a reader. Nevertheless, they are within reach of any Bachelor's degree-level student of physics and mathematics. The reader with a basic scientific background may benefit from a quick overview.

Chapter 2 presents the climatic variables and data. The large institutes and organizations (NOAA, GISS, CRU) make climatic data available. "Historic" temperatures overlap neatly from 1850 or 1880, eras when

7 Of course, it is important to have the necessary software, such as Matlab® or similar.

thermometric measurements started to become widespread around the world. Data from before this time can be qualified by paleoclimatology, and are given in the form of reconstructions, created using traces, markers or substitution measurements (proxies) left by the climate on the Earth and in the oceans. These two types of data (instruments and proxies) cannot be used in their raw form for the purposes of identification. They need to be linked together in order to be processed as a continuous data stream through time.

Chapter 3 discusses a regrettable debate, a *war of graphs* where parties exchange inconsistent arguments, often reduced to the display of climatic signal lines. At times, these lines are contested to the point where the credibility of paleoclimatic data in general is in doubt, thus jeopardizing the very principle behind the identification of a climatic model. In practice, we are able to totally disregard this controversy by processing all data available without taking one side or the other. This collection of data is not exhaustive, but its diversity is such that our conclusions cannot be accused of being obtained through "cherry picking".

Chapter 4 introduces the structure of the models which we would like to identify. This is taken from the class of models known as EBM, or Energy Balance Models. The simplest are static models, reduced to three or four coefficients. They are too basic and unable to give an accurate picture of the reality. The most complex models already constitute the first drafts of GCMs, or General Circulation Models, for the atmosphere or oceans. These have too many parameters and cannot be identified since there are many which are redundant in terms of input/output. The characteristic feature of the structure which is used is that each piece of input data is assigned a balance sensitivity coefficient, but all are subject to the same transients of heat transfer. A "black box" is created as a result, in which certain physical coefficients appear in the form of combinations and remain out of reach in their individually. This is the compromise made to find the right balance between too many and too few parameters.

Chapter 5 brings together the assumptions relating to fundamental climatic parameters of energy balance models, as well as their uncertainty ranges. These are taken, directly or indirectly, from official IPCC publications: SPM and AR5.

Chapter 6 examines the identification method. It is the simplest and most reliable possible, that of the least square of output error (OE: Output Error

Method). Given the nature of the data, the model is not statistically optimal, but there is no reason to assume that it is far from it. Moreover, as the uncertainty calculation does not result from an estimate of the *maximum likelihood*, a method to calculate the uncertainties is specifically developed – without giving rise to any particular difficulties. Finally, the end product is a reliable instrument, both in terms of determining parametric estimations and the associated uncertainties.

Chapter 7 gives a first overview of the results of identification. In our climatic archive catalogue, we have selected[8] an initial combination out of sixteen (four temperature reconstructions and the same amount for solar irradiance). Two identification methods are shown. The first is a "free" identification, whereby the six parameters of the structure chosen in Chapter 4 minimize the error between the overall simulated temperature and the historic temperature data, without any *a priori* constraints being imposed. The second is a "forced" identification, where some parameters are fixed to comply with the assumptions given in Chapter 5. Some of the parameters in the free identification are located very far outside the IPCC range, especially with regard to the climate's sensitivity to solar irradiance. As long as we are dealing with recent warming (end of the 20th Century), the visual examination of the simulated output temperature from both equally reproduce this warming. The difference being that with free identification, the contribution of solar irradiance is highly predominant over that of CO_2 levels, while the opposite is true for forced identification. However, the IPCC experts claim that it is physically impossible for solar irradiance to have a significant impact on the climate. It is therefore important to go further, and not simply rely on a visual impression. Statistical analysis starts by assessing the autocorrelation function of the output error, and the cross-correlation with input signals. This confirms that the constraints of forced correlation lead to a strong correlation with solar irradiance. This points towards a cause-effect relationship which is not taken into account. Even more importantly, areas of related uncertainty show that the IPCC's assumption of weak sensitivity to solar irradiance must be rejected. There is a very low likelihood of error in such a result. This rejection is based, not on considerations from theoretical physics, which are excluded from our study, but on the statistical processing of observations. If the observations and processing are correct, one must therefore conclude that the assumptions given above are false.

8 The one which, with hindsight, appeared to be the most representative.

Chapter 8 extends this analysis to the sixteen possible combinations among the four paleotemperatures and the four reconstructions of solar irradiance. The overwhelming majority of these confirm the previous analysis. The exceptions all arise from the same temperature reconstruction, that of Phil Jones and Michael Mann [JON 04], who are active protagonists in the *graph war* mentioned above. Even though processing their reconstruction does not allow us to reject the hypothesis of weak sensitivity to solar irradiance, it does not confirm it either. On the other hand, high sensitivity to solar activity, statistically validated, cannot be contested with the argument that the mechanisms of action are not accurately known. In terms of sensitivity to CO_2, the IPCC window is very wide, with the extreme values differing by a factor of six. Unfortunately, the nature of historic and paleoclimatic data is such that identification is unable to narrow this window. Instead, it moves the whole range downwards, reducing the highest, and seemingly most exaggerated, values. It cannot even be excluded that human activity has a negative impact on global temperatures.

Chapter 9 compares results of the IPCC simulation, over the historical period mentioned above (1850 to today), with simulations from the identified models. Temperature reconstructions observed are similar, but contributions of natural and anthropogenic factors are turned upside-down. It would appear that the human influence conclusion for global warming is predetermined by the mode of generation of the input data as defined by the IPCC.

Chapter 10 offers long-term climate projections. To do so, scenarios created by the IPCC itself are used. Representative Concentration Pathways, or RCPs, put forward a series of profiles for future CO_2 concentrations. Unsurprisingly, simulations with constraints are alarming, especially in the worst case scenario, the so-called "business as usual" (RCP 8.5). Projections which result from free identification are much less worrying. Only a minority of cases exceed the allegedly critical threshold of a temperature increase of two degrees above pre-industrial levels by 2100.

Chapter 11 uses the short-term predictive capabilities (from several years to several decades) of the energy balance models. The climate models identified can be converted into state space models (in the systems theory sense). Observations available presently can be used to estimate this state through traditional techniques (Kalman filtering), from which it is then possible to carry out a prediction on the future climate using the current state estimated. The state in question includes both the climatic process itself and the input signals, especially those resulting from solar activity. The initial predictions do not provide any spectacular results. With natural climate variability, it would be necessary to wait another decade or so to be able to understand whether the current stagnation of temperatures is only a pause in warming, or the prelude to an out-and-out turn around. Nevertheless, looking back to the situation in 2000 when nobody would have predicted a possible slow-down in global warming, blind simulations carried out without including any information following 2000, predict with surprising accuracy the temperature levels that we are witnessing today.

Chapter 12 concludes that, on the basis of climatic observations, it is not possible to prove that human activity has a significant influence on the climate, and everything indicates that the determining factor is solar activity. Natural climate variability must be added to it. The paroxystic warming of the last quarter of the 20th Century resulted from the conjunction of strong solar activity with a loss of heat from the oceans under the effect of erratic deep ocean circulation. The ongoing steady values result from the inversion of these two causes. The IPCC's position is exactly the opposite, basing their arguments on physical and mathematical models. However, as yet, all predictions based on the IPCC models have been contradicted by later observations, while the model identified, using observations which go back more than a millennium, is confirmed by the current evolution of the climate.

2

Climatic Data

2.1. Sources

A number of organizations or institutions publish their climatic data. The following are the most well-known:

– Hadley Center and Climate Research Unit (CRU, of East Anglia University).

– Goddard Institute for Space Studies (GISS), of NASA.

– National Climatic Data Center (NCDC), of the National Oceanic and Atmospheric Administration (NOAA).

– World Data Center for Paleoclimatology, (Boulder, Colorado), which is also part of the NOAA.

Other bodies are more specialized, such as: the IRMB (Institut Royal Météorologique de Belgique), PMOD (Physikalisch-Meteorologisches Observatorium, Davos), LASP (Atmospheric and Space Physics at the University of Colorado), CDIAC (Carbon Dioxide Information Analysis Center), and so on.

We have not attempted to obtain all potentially available information, by directly contacting authors, publishers, or institutes. Our objective is not to be exhaustive, but to have a catalogue of climatic data which is representative in its diversity.

2.2. Global temperature

2.2.1. *Modern temperatures*

Global temperatures are collected, processed and updated by the large organizations mentioned above. We chose the data set HadCRUT4 of the Hadley-Center-CRU [MOR 12] for the simple reason that it starts in 1850, whereas those of GISS and NCDC don't start until 1880. In any case, the databases overlap well with each other. The average annual global temperatures of Earth surfaces and oceans, over a given year *t*, are noted as $T_G(t)$[1].

Up until the last few years, when satellites have partially taken over, global temperatures were measured under a shelter, at 1.50 m above the ground, and ocean temperatures were taken from the surface of the water, at a depth which varies depending on the depth of the cooling water intake on ships' engines (a few meters). The global average may not be foolproof, but it is nevertheless an essential indicator.

Since the end of the last great glacial period 8,000 years ago, global temperatures have not varied by more than one or two degrees. Climatologists measure variations with respect to a reference temperature, which is more practical than using 0°C. By convention, global temperature T_G is defined as an anomaly[2] of global temperature, assessed in relation to its average over the period 1850-1900.

There are other reference periods, depending on organizations and requirements: 1961-1990 (Hadley) and 1951-1980 (GISS, NCDC). When the IPCC presents its future temperature predictions, they are sometimes in relation to a "current temperature" based on the average of the period 1986-2005 (see Figure 9.4).

When processing data, we will do it from raw values of annual temperatures (shown by thin lines on Figure 2.1). For visualization purposes, they can be smoothed (bold lines), either by moving averages, or by Gaussian or binomial filters. It is preferable to use spline functions with penalties on derivatives (Matlab®'s *csasp* tool), which work without data

1 In this year $t =$ 2014 at the time of writing, $T_G(2014)$ is as yet unknown.

2 Anomaly, without the connotation of abnormality.

truncation at the beginning or end of the period and without introducing a significant edge effect. The smoothing parameter p is fixed below at $p = 0.1$.

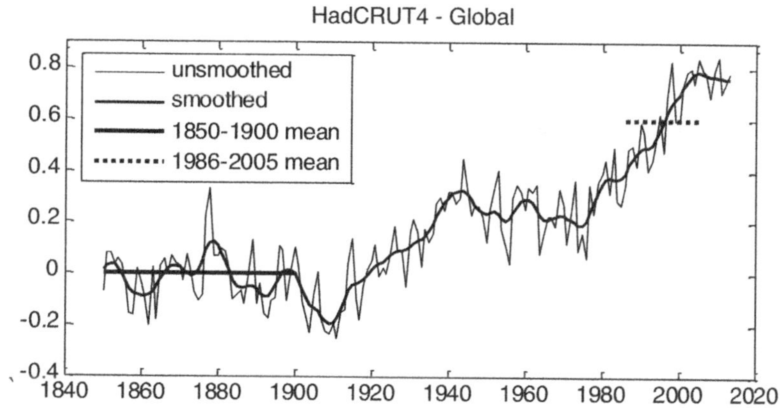

Figure 2.1. *Global temperature, unsmoothed and smoothed curves.*

2.2.2. *Pre-industrial temperature*

The pre-industrial equilibrium temperature T_E is that which should have been observed when CO_2 was close to its pre-industrial levels (before 1750), assuming that solar and volcanic activity were stable at their theoretical averages. Since these conditions have never been fulfilled and measuring instruments not being available at that time, the hypothetical temperature T_E is based on theory.

Since this temperature T_E cannot be fully relied upon, a reference is defined as the average temperature over a plausible representative period, starting later than 1850 (temperature measurements beforehand are too uncertain). The period 1880–1909 was used, but in the last IPCC report, the period 1850–1900 is used most often, and this is the period which will be referred to throughout this work.

Under this convention, the pre-industrial temperature T_E remains unknown, but it is presumed to be close to zero. Whatever estimations are made afterwards, when global warming is addressed, it will be assessed with respect to the average temperature over the period 1850–1900.

2.2.3. *Paleotemperatures*

Temperature reconstructions from before 1850 are based on archive material: traces left on the planet by the climate in the past. Measurements of substitution, or proxies, are extracted from this data. One of the most well-known examples is tree rings, whose thickness varies depending on climatic conditions. Others are based on oxygen-18 isotope (^{18}O), whose ratio with ^{16}O in precipitation varies according to climatic conditions. Water from rain or snowfall is either stored in ice caps or trapped in the form of calcium carbonate found in corals, foraminifera, lacustrine or maritime sedimentary layers, stalagmite deposits, and so on. The value of such a proxy is that the relation between the ratio of ^{18}O and ^{16}O isotopes and temperature is now well established. The "isotopic thermometer" is now well calibrated, but dating remains a source for errors. As a result, the isotopic proxies are renowned for their good resolution at low frequency (high level of accuracy on the average value over a long period). Although tree rings more faithfully track annual temperature variations, they are not as good at recording gradual trends. There are a number of proxies: the profile of temperatures measured in boreholes can be used to reconstruct past temperatures by inversing the laws of thermal conduction, archives recording harvest start dates, function of summertime climatic conditions, and so on. The difficulty comes from selecting the proxies, merging them, calibrating them and aligning them with modern measurements.

Figure 2.2 [MOB 08] shows a sample of eleven of the proxies used, amongst others, by Moberg and Loehle to build their respective reconstructions. Many of them confirm the existence of a medieval warm period (MWP, between 800 and 1200), and show that the little ice age (LIA, between 1600–1850) was not an event limited only to central England. Some proxies can even be used to visually assess the temperature deviations from minimum to maximum to within around 1°C. Understandably, final reconstructions can differ markedly, depending on the selection and statistical processing of proxies.

Of the climate archives from which we were able to gather data, we kept four among those which went back further than the year 1000. We took on no more data so as to prevent the number of combinations with reconstructions of solar activity from becoming unmanageable. Nevertheless, these four paleotemperatures (Figure 2.3) constitute a relatively representative collection of the variety of existing reconstructions: Moberg (2007), Ljungqvist (2009), Loehle (2007), Jones & Mann (2004).

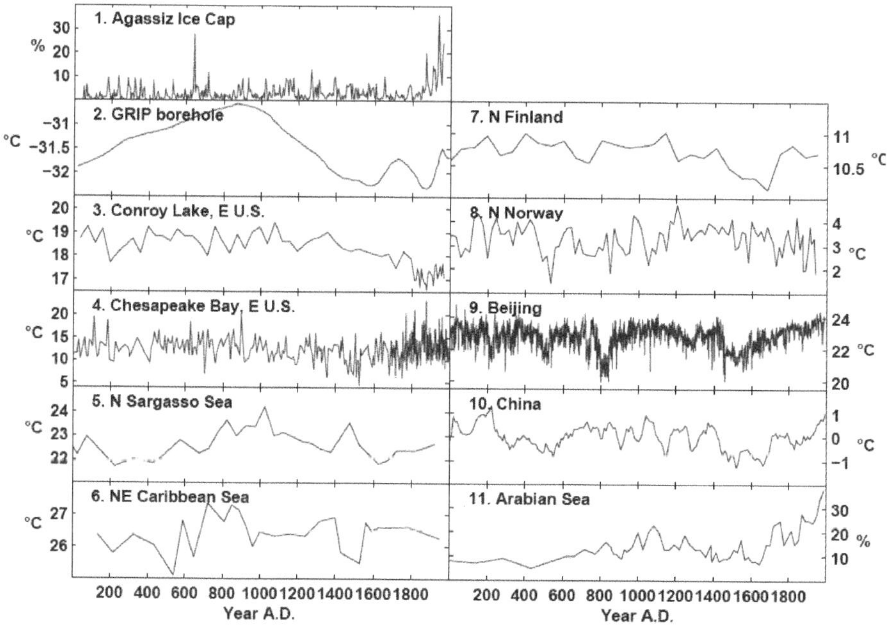

Figure 2.2. *Eleven low frequency proxies*

When the AR5 was released at the end of 2013, among the 15 paleotemperatures shown in Figure 5.7 of the AR5, the first three from the above list were featured. The Jones and Mann data is missing. However, there are four more recent reconstructions from M. Mann, as well as a reconstruction from Christiansen and Ljungqvist [CHR 12], which we disregarded due to problems matching its data with historical measurements. Despite its absence from the AR5 catalogue, the Jones and Mann data [JON 04] will be retained because of its closeness to the hockey stick graph [MAN 99], which we would have kept for its "historic" interest had it not been limited to the second millennium.

In any case, this was the choice that we made, dictated to some extent by chance and circumstances, and we believe that it gives a relatively representative depiction of the variety of reconstructions available.

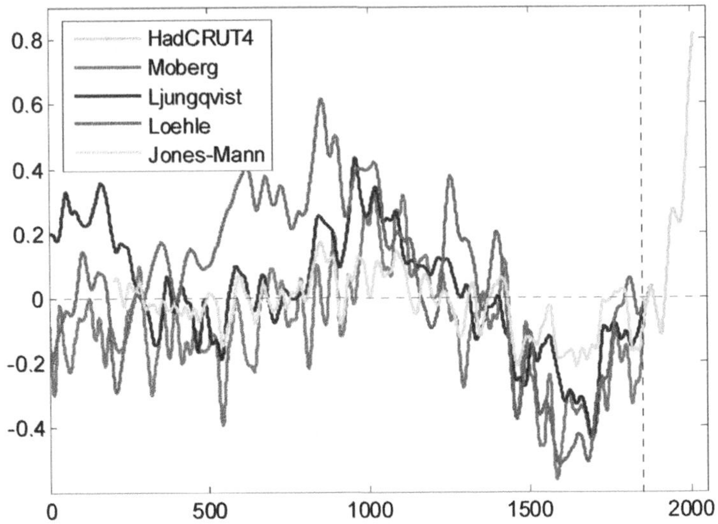

Figure 2.3. *Four temperature reconstructions, collated in HadCRUT4*

In order to obtain a set of single piece data we spliced them with the global temperature HadCRUT4, from 1850, after alignment by equalizing the average values over the overlap period. This was done without recalibration, maintaining the original scaling factors, as shown in Figure 2.4. It can be seen that the quality of the reproduction of thermometric temperatures by reconstructions is far from perfect.

Some of these reconstructions are supposed to be global, others are limited to continents or to the northern hemisphere. However, they are all linked without distinction to the same global temperature (HadCRUT4), since uncertainty caused by proxies appears to be greater than the differences that can be seen over the historic period, between continents and oceans, or between the northern and southern hemispheres.

The overall view of these curves (Figure 2.3), over more than a millennium, confirms that they come from the same climatic history, and that the frequent concordance in their variations is not down to chance. However, it can be noted that the relationship with the high temperature excursions between the PCM and the PAG varies by a factor of three, between the Jones-Mann and Loehle reconstructions. It is not our aim to decide which is the most representative of the actual evolution of global temperature.

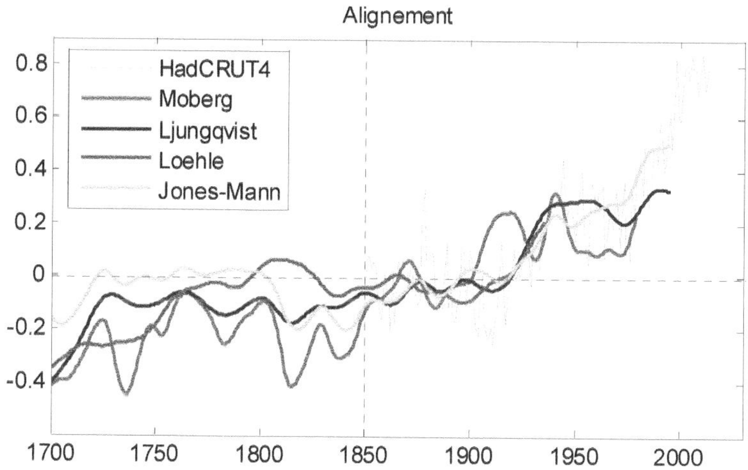

Figure 2.4. *Linking to HadCRUT4*

2.3. Concentration of CO_2 in the atmosphere

$C(t)$ is the concentration of CO_2 in the atmosphere, expressed in parts per million by volume (ppm). The line in Figure 2.5 links three segments:

– With a lack of data from before the year 1000, we can assume that the concentration is equal to a nominal pre-industrial level C_E, of 280 ppm.

– Between 1010 and 1958, concentrations are taken from the air bubbles trapped in Antarctic ice caps in Lawdome[3].

– Since 1959, atmospheric concentrations are those measured from the center at Mauna Loa (Hawaii)[4].

3 http://cdiac.ornl.gov/ftp/trends/co2/lawdome.smoothed.yr75.

4 ftp://ftp.cmdl.noaa.gov/ccg/co2/trends/co2_annmean_mlo.txt.

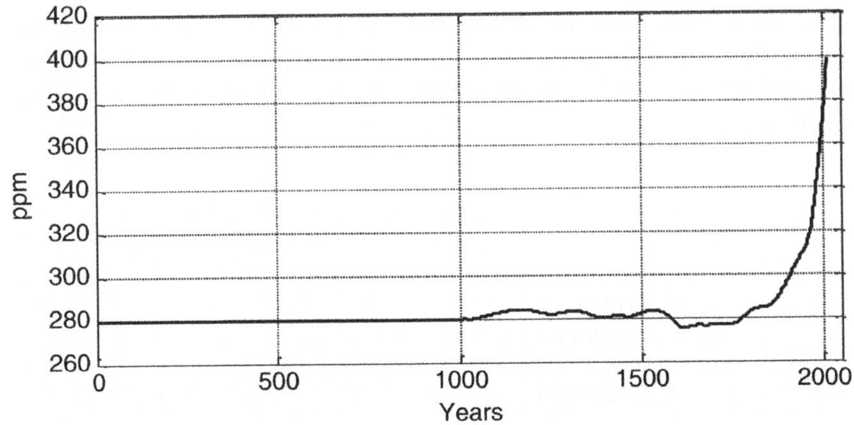

Figure 2.5. *Concentration of CO_2, in parts per million (ppm)*

According to Arrhenius (1896), confirmed by Myrhe (1998), the CO_2 in the atmosphere acts on the climate in relation to the logarithm of its concentration. This is why a non-dimensional variable u_1 is defined (indices 2 and 3 respectively refer to solar irradiance and volcanic activity):

$$u_1 = \log(C / C_E) \sim \frac{C - C_E}{C_E}$$

Hence $u_1 = 0$ for the pre-industrial value. At each doubling of CO_2, u_1 increases by $\log(2) = 0.6931$. After CO_2 has quadrupled, $u_1 = 1.3862$.

2.4. Solar activity

The fluctuations in solar activity constitute a second potential cause for climatic variations. This has a number of different effects.

Total irradiance

A first manifestation of solar activity is the total energy flow intercepted, at Earth's level, in W / m^{-2} (TSI : Total Solar Irradiance). In fact, the solar constant, of around 1366 Wm^{-2}, varies little (variations of around ‰). The first satellite measurements date from 1978. Since satellites have a limited lifespan (between 10 and 15 years), the results they give are scattered as in the top part of Figure 2.6[5].

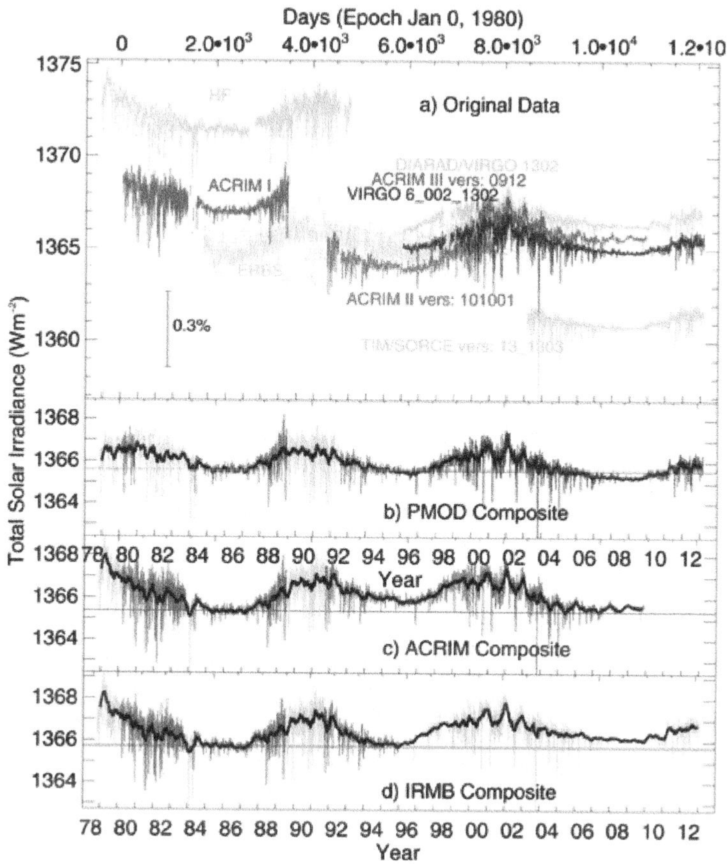

Figure 2.6. *Satellite measurements of TSI (source PMOD)*

Lines (b), (c) and (d) show different composite measurements which are extracted from them. It is easy to distinguish the succession of Schwabe solar cycles, over 11 year periods, numbered 21, 22, 23 and 24. The three composites do not overlap well: starting from the same base in 1986, the

5 Source: World Radiation Center website, from the PMOD: http://www.pmodwrc.ch/pmod. php?topic=tsi/composite/SolarConstant.

difference between *IRBM* and *PMOD* reaches 0.6 Wm^{-2}, which is a considerable figure, given the peak amplitudes of the order of 1 Wm^{-2}.

There are also measurements of spectral distribution IR/UV of total irradiance, which could have an influence on the climate. In addition, it is important to briefly highlight that since satellites are not able to measure the overall radiated flow to better than one percent accuracy while the sun is a quasi-punctual source, they are even less able to measure with any accuracy the energy reflected or re-emitted in the visible spectrum or infra-red as this flow is distributed over the surface of the Earth. The numerical evaluations of net radiation that can be found (Trenberth, 2009), accurate to a tenth of a Wm^{-2}, can never result from direct measurements.

Wolf Number

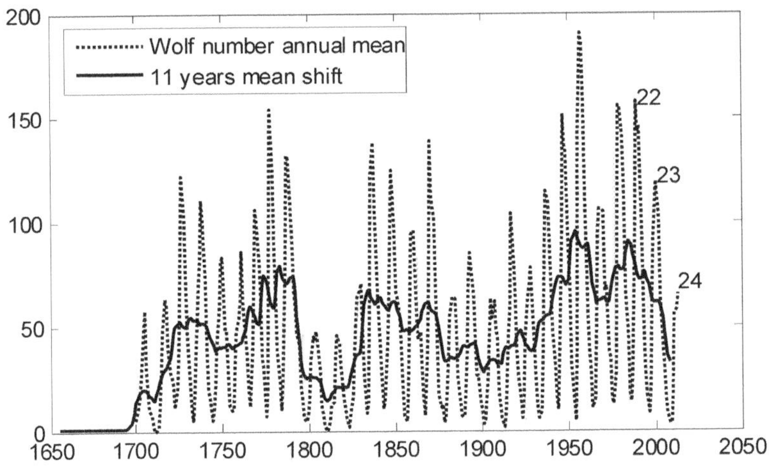

Figure 2.7. *Sun spots*

A second effect of solar activity is the number of sun spots observed, according to the counting method standardized by the astronomer Rudolf Wolf in 1855. This number or *Wolf index* is written: $R = k\,(10\,g + f)$, where f is the number of spots visible, brought together into g groups, k denotes a coefficient (generally <1) depending on the quality of the images. The spots have been observed since 1610, the date of the invention of the telescope accredited to Galileo. Between 1650 and 1700, these spots were rare

(Maunder minimum) and sporadic. The Wolf index is well reconstituted from 1700, and even more so from 1750. Wolf started numbering the cycles in 1750. The peak of cycle 24 was reached in 2012–2013.

Figure 2.7[6] shows the evolution of the Wolf number since 1700, and its moving average over 11 years. This clearly shows that average solar activity reached its maximum values during the last half-century and is now in decline.

Isotopic proxies

A third effect of activity is the solar magnetic field. Direct measurements are limited to the recent period. However, an alternative measurement over a millennium is calculated with levels of isotopes known as *cosmogenic*, or *cosmonuclides*: beryllium-10 (^{10}Be), or carbon-14 (^{14}C). These isotopes are generated in the atmosphere by cosmic rays, which are themselves regulated by the solar magnetic field.

Two databases were used (Figure 2.8), based on ^{10}Be (changed sign in order to respect TSI direction of variation).

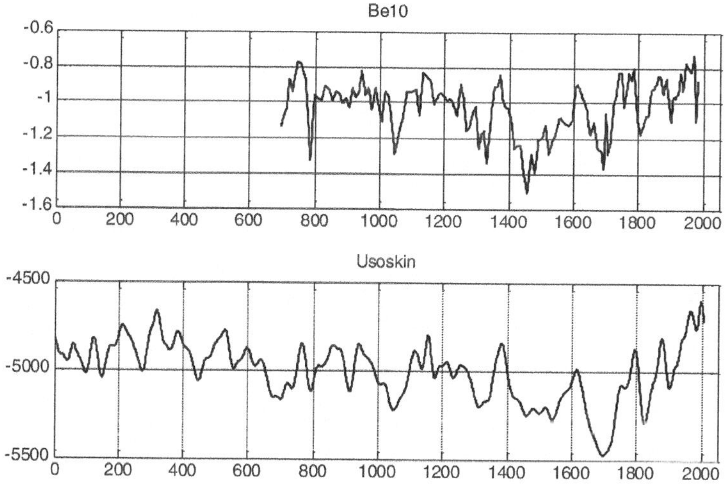

Figure 2.8. *Proxies for TSI*

6 Source: ROYAL OBSERVATORY OF BELGIUM. Av. Circulaire, 3- B-1180 BRUSSELS http://sidc.oma.be/html/sunspot.html.

The first graph, from Delaygue and Bard [DEL 10] starts in 695, while the second is from Usoskin *et al.* [USO 08]. They both correct the raw data taken from ice caps in order to take various factors into account, such as the conditions of atmospheric transfer, the deposit process, or terrestrial geomagnetism. The units are different, but this is irrelevant since the proxies will be converted into TSI units (Wm^{-2}).

Partial reconstruction

The three types of indicators of solar activity stated above are too strongly correlated for an identification algorithm to be able to assess precisely what can attributed to one or another in global temperature variations. Inversely, they are not close enough to allow us to splice them together without difficulty.

Figure 2.9 shows four reconstructions spanning four centuries, (period 1610-2013), splicing sun spots (Figure 2.7) with modern measurements (Figure 2.6).

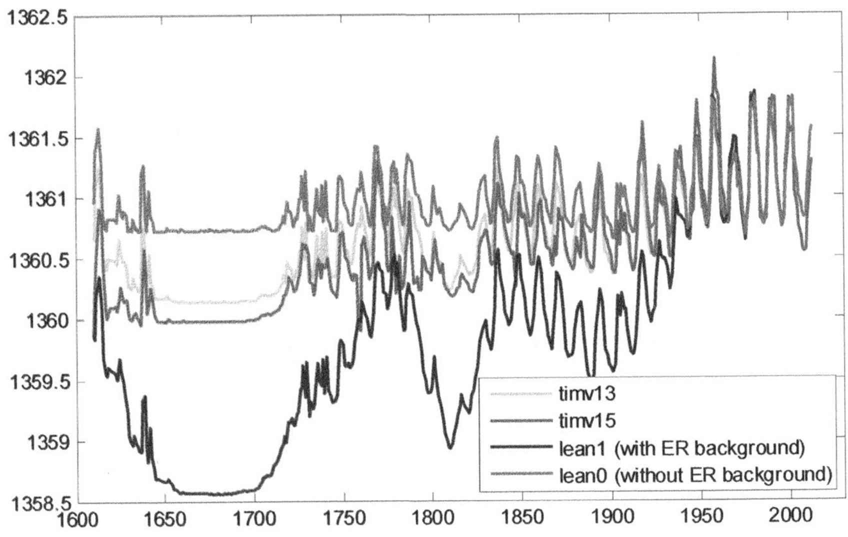

Figure 2.9. *Reconstructions over four centuries (1610–2013)*

Lines labelled lean0 and lean1 come from the reference [LEA 04]. Lines timv13 and timv15 were determined by G. Kopp [KOP 05] using PMOD composites, respectively based on versions TIM V13 and TIM V15 of TIM/SORCE (Total Irradiance Monitor, system fitted on the NASA Solar Radiation and Climate Experiment satellite).

The differences are great. The line labelled lean0 is almost proportional to the annual average number of sun spots. The other three include corrections known as Ephemeral Region Background[7]. They also rely on the solar models established by comparison with stars similar to our sun. The authority on the subject is Judith Lean, leader of the IPCC experts on solar matters, and principal or secondary author of the four reconstructions above. Ultimately, these corrections amplify the low frequency components of the Wolf index, with ratios ranging from 1 (for lean 0) to 5 (for lean1). Therefore, this represents a dynamic linear correction rather than a non-linear algebraic correction.

Furthermore, another reconstruction covering four centuries can also be referred to (Figure 2.10), which is not included in Figure 2.9, the data from which we have not collected. It comes from Shapiro *et al.* [SHA 11] and enhances the ER background effect almost two-fold compared to lean1, under the name of solar modulation potential.

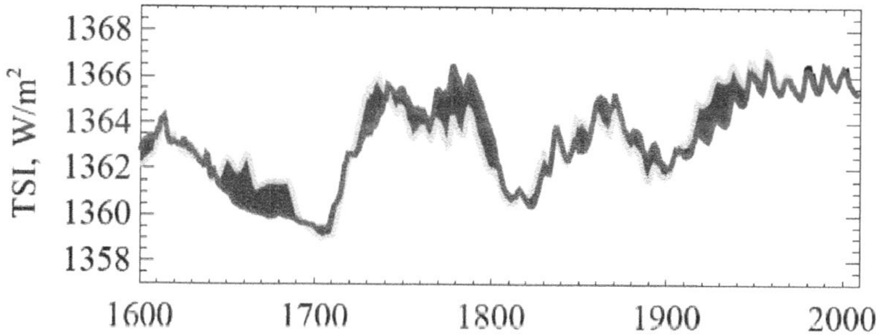

Figure 2.10. *Total irradiance reconstructed by Shapiro*

7 The "ephemeral regions", discovered in 1973, are small magnetic dipoles which disappear before they are able to turn into spots, but the activity of which has an effect on both the TSI and the solar magnetic field.

Finally, reconstructions covering several millennia are available, which are based on isotopic proxies (Figure 2.11). Unfortunately, they are not linked to satellite measurements, which means they we are not able to use them. Moreover, their scales are incompatible: visibly, they have been calibrated with different four-century TSIs.

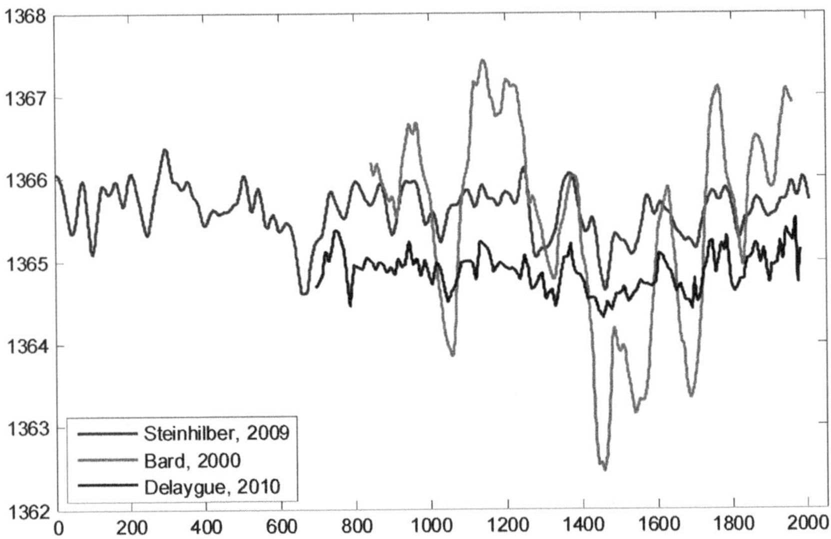

Figure 2.11. *Irradiance reconstructions over millennia*

The Bard line could conveniently align with lean1, the others with *timv13* or *timv15*, but it would be inappropriate to recalibrate this data behind the authors' backs. However, cosmogenic proxies, like those in Figure 2.8 should, by their very nature, be calibrated with TSI observations or the partial reconstructions in Figure 2.9. We are therefore completely free to do so.

A selection

The correlation of all the solar activity indicators (TSI, spots, magnetic field, cosmogenic proxies) justifies them being merged into a single representation. The TSI is adopted as a common reference, although the energetic factor alone does not appear to be sufficient to take into account all the climatic variations caused by solar activity. In practice, we link the two proxies in Figure 2.8 to the following two composite lines: firstly, lean0, for its proportionality to sun spots, and secondly timv15, which is the most recent (2014). In this way, we are able to obtain the four reconstructions given in Figure 2.12, in the graphs on the left.

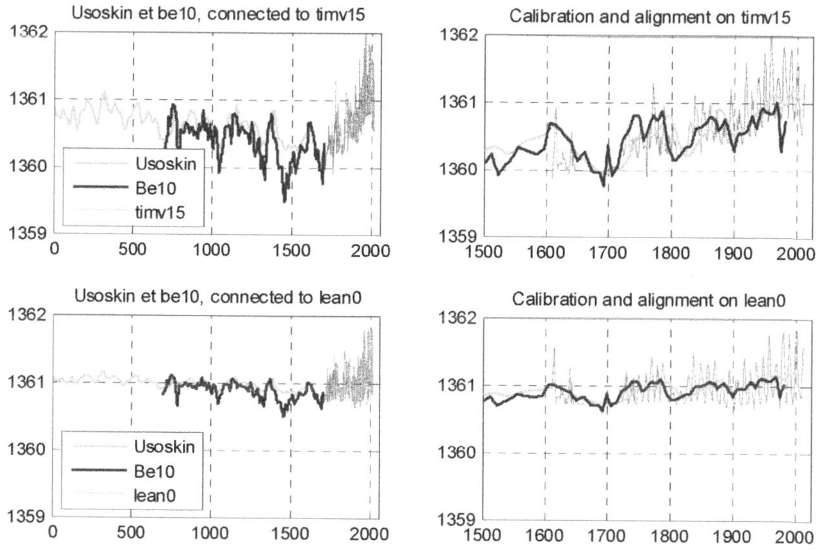

Figure 2.12. *Four irradiance reconstructions*

In order to carry out the splicing, we determine by least squares, over the period of overlap, an affine transformation $y = ax + b$ between reference y (lean0 or timv15 as appropriate) and the series x to be connected. Subsequently, we apply the relation obtained for the whole series x, and connect from 1700 onwards. The graphs on the right in Figure 2.12 highlight the quality of the connection, significantly better for irradiance than they are for temperatures (see Figure 2.4). Other options could have been considered; the analysis of principle components, for instance, or possibly using non-centered PCAs. Fortunately, the quality of the overlap is good enough so that the technique used only has a minimal impact on the result.

Depending on the option (lean0 or timv15), the amplitude of the excursions ranges from around one to three. As a result, it is not surprising to identify solar sensitivity coefficients which have the same ratio, according to the reconstruction adopted.

Finally, in order to define the input signal u_2 of the climatic process to identify, TSI anomalies are considered in relation to the respective average values U_E of each of them. Furthermore, a factor of 1/4 is introduced, in relation to the ratio of the area of the cross-section to that of the sphere. Net irradiance can then be obtained by taking into account the nominal terrestrial albedo (0.3) with the co-factor 0.7:

$$u_2 = (U - U_E) \times \frac{1}{4} \times 0.7 \quad (\text{in } Wm^{-2})$$

The anomalies for net irradiance are of the order of a tenth of a W/m^2. This will again appear extremely minor compared to the other energy flows. Clearly, this does not exclude the possibility that solar activity variations, for which we used TSI as an indicator, may have an effect on the climate by other mechanisms.

2.5. Volcanic activity

Volcanic activity affects the climate through the production of aerosols (ash, sulphates). The sun's rays are obscured by these aerosols and partially reflected back into space, having a negative effect on temperatures. Aerosols also have an indirect effect by contributing to the formation of clouds in a complex way depending on the chemical composition and the dimensions of the particles. The archives of the Antarctic or Greenland ICE can be used to calculate the optical depth[8] of aerosols caused by eruptions (AOD:Aerosol Optical Depth), year after year, as well as the dimension of the volcanic particles. In an identification context, where the aim is not to reconstitute the corresponding radiative forcing by physical rules, it is good enough for us to use an indicator, and the optical depth is perfectly suitable. NOAA released a reconstruction of optical depth over the period 800-2000, designed by T.J. Crowley and M.B. Unterman [CRO 13].

For input u_3 of the model, we will take the difference of the AOD with respect to the average value over the period 800-2000, with a changed sign so as to create positive sensitivity. Volcanic activity which is temporarily zero is therefore negative in relation to the average which is thus represented

8 Optical depth depends on wavelength (here, 550 nm). For weak values, optical depth equals a fraction of the rays obscured.

as a positive anomaly. In the absence of past or future data, the anomaly is extrapolated by zero. In Figure 2.13 it is possible to discern the last three high impact eruptions: Agung (1963), El Chinchon, (1982) and Pinatubo (1991).

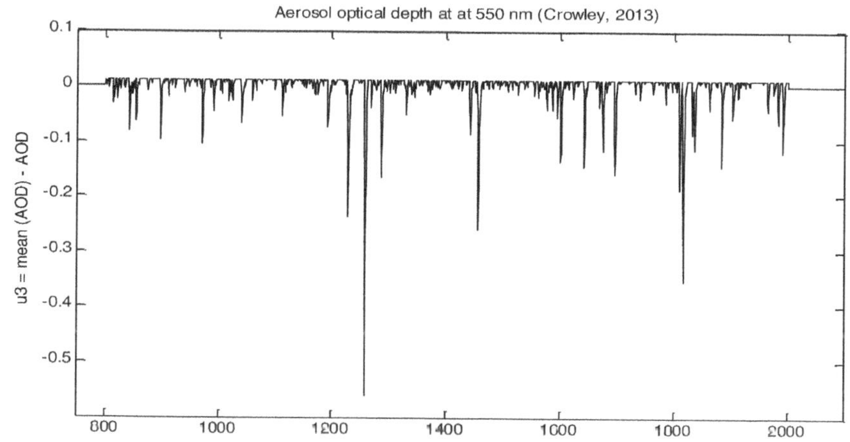

Figure 2.13. *Indicator of volcanic activity*

3

The War of the Graphs

Who controls the past controls the future.
Who controls the present controls the past.
Orwell, 1984.

Such a chapter should not be found in a scientific work. It is necessary, however, given the controversial context unsettling the climate change issue.

3.1. History

Since its inception, the IPCC heavily communicated on the link between global warming and the exponential increase in the concentration of CO_2 in the atmosphere. The argument reached its climax following the release of the third IPCC report in 2001, with the publication of the temperature curve known as the hockey stick graph by Michael Mann *et al.* (1998, 1999). This famous curve was used as the background to a number of climate conferences. Figure 3.1 below, shows Sir John T. Hougthon, founding member of the IPCC, during a speech on climate change (High Wycombe, 2005).

The hockey stick graph was an intensively used weapon in what some people referred to as 'the battle of the graphs'. As early as 1995, Jonathan Overpeck (Director of the Institute for the Study of Planet Earth), reportedly argued that the Medieval Warm Period had to be "gotten rid of"[1]. Perhaps he was thinking of the Lamb reconstruction, published in the first IPCC evaluation report in 1991 (Figure 3.2)[2].

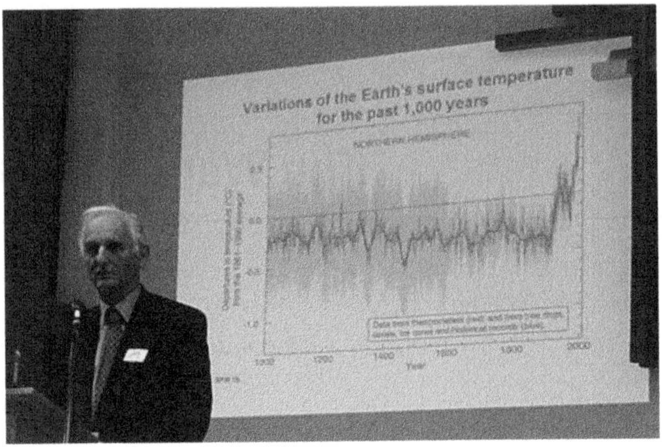

Figure 3.1. *The hockey stick graph*

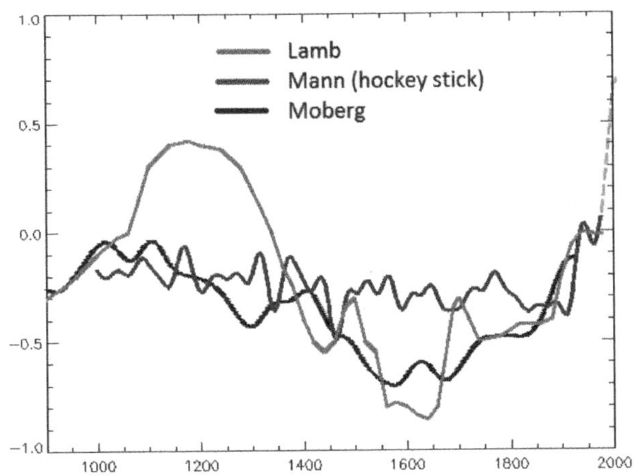

Figure 3.2. *A "plate of spaghetti", including the hockey stick curve*

1 David Deming's witness statement before the American Senate: "I received an astonishing email from a major researcher in the area of climate change. He said, "We have to get rid of the Medieval Warm Period" http://www.epw.senate.gov/hearing_statements.cfm?id= 66543.
2 Source: http ://en.wikipedia.org/wiki/Intergovernmental_Panel_on_Climate_Change http://en.wikipedia.org/wiki/File:Ipcc7.1-mann-moberg-manley.png.

When questioned, Overpeck did not confirm or deny whether he had made such a statement, but he did write the email[3] with the subject "The new "warm period myths"", addressed to Keith Briffa in 2005: "I get the sense that I am not the only one who would like to deal a mortal blow to the misuse of supposed warm period terms and myths in the literature....pure rubbish". He also asks whether Mr. Briffa has any "cool" figures in this regard. Mr. Briffa is indeed the author of a reconstruction, made famous after being truncated by Phil Jones (Director of the *CRU*), in order to "hide the decline"[4]. Indeed, from the 1960s, proxies have indicated a decline, which is incompatible with the increase in thermometric temperatures.

Michael Mann has a similar mindset, writing to Phil Jones in 2003, and copying in other luminaries of the climatological community (including Overpeck): "It would be nice to try to "contain" the putative "Medieval Warm Period""[5]. In the Jones & Mann reconstruction included in our datasets, the medieval peak is not completely hidden, as it was in the original hockey stick graph, but it has indeed been extremely well "contained".

On the other hand, the Loehle reconstruction shows the highest medieval temperatures in our collection, up to 0.3 or 0.4°C higher than in the Jones & Mann reconstruction, though without reaching the levels of the Lamb graph. Loehle, member of the Society of American Foresters and an expert in his field, preferred to eliminate the proxies given by tree rings, which were given pride of place by Jones and Mann, and are known for their tendency to reduce large-scale climatic variations. It was harshly received: "... The Loehle paper is awful as you know..."[6]. In AR5, (Chapter 5: Information from Paleoclimate Archives), Table 5.1 nevertheless includes this reconstruction, with the warning "almost no tree rings".

3 These emails were not intended to be made public. But since "Climategate", they could no longer be ignored.

4 Email from 16/11/99, from Phil Jones to: R. Bradley, M. Mann, M. Hughes Cc: K. Briffa, T. Osborn: "... I've just completed Mike's Nature trick of adding in the real temps to each series for the last 20 years (ie from 1981 onwards) and from 1961 for Keith's to hide the decline....".

5 Email from 04/06/2003, from Michael Mann to Phil Jones, R. Bradley, Keith Briffa, Tom Crowley, Michael Oppenheimer, Jonathan Overpeck, Kevin Trenberth, Tom Wigley.

6 Email from 05/12/2007, from Phil Jones to Michael Mann.

However, this Chapter 5 in question [MAS 13] does not make the slightest mention of the famous publication from M. Mann *et al.* [MAN 99] neither in the text nor among the some 1,000 specific bibliographical references in this chapter. Given the extensive use that the IPCC made of it in the past (cited six times in the Third Assessment Report), and the controversy it still causes, this absence is peculiar[7].

Moreover, there is a recurrent debate on the local or global nature of large climatic events. Figure 2.2 shows that proxies which detect a MWP or a LIA, are distributed across the Earth's surface, including the oceans and are not limited to central England or to Greenland (green-land) mythologized by Eric the Red. The IPCC admits that in the last SPM (section 5.5):

"Continental-scale temperature reconstructions show, with high confidence, multi-decadal periods during the Medieval Climate Anomaly (year 950 to 1250) that were in some regions as warm as in the late 20th Century. These regional warm periods did not occur as coherently across regions as the warming in the late 20th Century (high confidence)."

This lack of synchronization is apparent in Figure 3.2. It is insufficiently emphasized at the whim of some governmental delegations[8], but on the other hand, it shows that the warm medieval period stretches over several centuries, while the sudden current warming only spans several decades. This lack of coherence in the reconstructions based on proxies is just as present for the modern-day (see Figure 2.4) as it is for the past.

The use of the expression "Medieval Climate Anomaly" can be seen above. In AR4, *Medieval Warm Period* was still also often used (although always in inverted commas). In the AR5, it has quite simply been removed from climate-specific language, to be replaced by the milder MCA: Medieval Climate Anomaly. This is the mortal blow that Overpeck was looking to deal regarding the use of this term.

---

7 According to *Google Scholar*, Mann (1999) is quoted 1681 times in scientific literature, Moberg 1008 times, Ljungqvist 103 times and Loehle 56 times (for their respective reconstructions).

8 Concerning text on the Medieval Climate Anomaly, Belgium and Ireland underscored that this phenomenon was regional in nature, unlike global warming in late 20th Century. (Kosolapova, E, 2013)

The war of the graphs is not limited solely to temperature reconstructions. It is also played out in the field of solar irradiance reconstructions. It is known that the IPCC recognizes the contribution of solar activity, but only in the reduced form of net TSI. This goes some way to explaining the medieval warm period, without contributing too heavily towards the 20th Century warming, which would harm the CO_2-centred hypothesis if its contribution were greater. With this in mind, anything that can emphasize medieval solar activity and reduce its significance in the 20th Century is very welcome. Therefore, Delaygue and Bard [DEL 10] note that "[their] results suggest that the recent solar activity is not exceptionally high for the last millennium". Indeed, we discover (Figure 2.11) that between 740 and 750 a local maximum briefly reached 20th Century values[9]. More overtly, in an EBM simulation study, T. Crowley [CRO 00] announced that he had completed a set of two reconstructions of irradiances by "an estimation of ^{14}C using fluctuations in ^{10}Be. The justification of the inclusion of this latter index is that none of the first two datasets leads to a medieval solar maximum which is comparable to the current day". Justification or merely "cherry picking"?

It is also known that the IPCC systematically refers to the date 1750 as representative of the pre-industrial era. By chance (?), this date is right in the middle of a local maximum reading of solar activity, between the minimums of Maunder (1700) and Dalton (1800). The variation shown by radiative forcing compared to the modern era is thereby significantly reduced. The effect is emphasized further by adjusting the calculation using the closest cyclical minimums to 1750 and 2011, claiming that these minimums are the most "stable". Thus, the IPCC can produce a variation of radiative forcing since the pre-industrial era of between 0 and 0.1 Wm^{-2}, reduced further by the AR4 report, where the figure was located between 0 and 0.3 Wm^{-2}. A glance at the Shapiro reconstruction (Figure 2.10) therefore leaves the reader puzzled, with min/max variations reaching 6 Wm^{-2}. Given the factor 0.7/4 to reach net irradiance, the result is variations in solar radiative forcing of over 1 Wm^{-2}. This distortion leads the IPCC to place question marks over this reconstruction (AR5, sections 2.2.1.2, 2.2.1.3 and 5.3.5.3[10]).

9 ER background correction also has a contribution to make on this issue.

10 "Simulations with an EMIC using a much stronger solar forcing (0.44% TSI increase from LMM to present, Shapiro *et al.*, 2011) appear to be incompatible with most temperature reconstructions (Feulner, 2011)".

Even modern TSI measurements are not safe from controversy. In a letter (16 September 2008)[11], addressed to his assistant Nicola Scafetta[12], Richard Willson, in charge of the satellite measurement system ACRIM (Active Cavity Radiometer Irradiance Monitor), complained that Frohlich and Lean [FRO 04] had made unjustified modifications to their TSI measurements (Figure 3.3, taken from AR4).

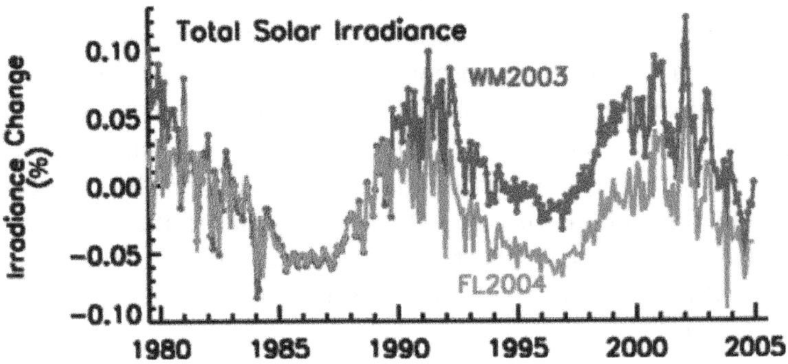

Figure 3.3. *TSI by Willson and Mordlvinov (2003) and correction by Fröhlich and Lean (2004)*

The correction cancels out the increase seen since 1986, and also reduces the possible influence of solar activity on the warming in the latter part of the 20th Century. Conversely, taking into account the time constant of the climatic system, it could contribute to explaining the stagnation witnessed after 2000. Furthermore, it could completely alter the evaluation of the ER background correction, or Shapiro's potential modulation.

11 Source: http://climatechange.thinkaboutit.eu/think4/post/judithgate_ipcc_consensus_was_only_one_solar_physicist/.

12 N. Scafetta studied at the University of Pisa – like Galileo. In 2014, he was published in the second edition of the journal *Pattern Recognition in Physics* (Copernicus publications!). This paper placed the Sun at the center of the climate system and concludes with a "doubt on the continuation and even the future acceleration of the warming claimed by IPCC". The publisher (Martin Rasmussen) was horrified to have allowed such heresy and took the drastic decision to take the journal out of existence. His report is stunning: http://www.pattern-recognition-in-physics.net/.

3.2. Inconsistent controversies

Observational data is very striking because they can be converted into graphs, which are more powerful than long speeches, and which the IPCC does not deign to call upon. Nevertheless, for the IPCC, the very high levels of confidence in the human influence on climate warming relies on physical knowledge. Conversely, historic data constitutes the basis of our work. This makes it more sensitive to the diversity of paleoclimatic reconstructions than other studies.

Figure 3.4 shows, on one side, the original M. Mann reconstruction (graph a), and on the other side, the Moberg reconstruction (graph b). The curves entitled hockey stick and boomerang were not drawn free-hand, but come from pure mathematical curve smoothing.

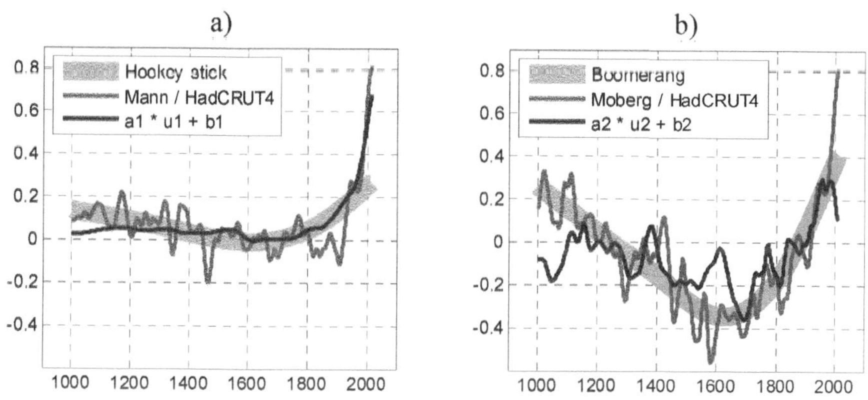

Figure 3.4. *Hockey stick or boomerang?*

After calibration and alignment by linear regression, the dotted lines reproduce respectively the reconstructions of log-concentration of CO_2 (a) and of solar irradiance (b). Obviously, the conclusions reached differ depending on the paleotemperatures selected.

The coefficients a_1 and a_2 can be assimilated to sensitivities, respectively to CO_2 and solar irradiance. In the "100% CO_2" hypothesis suggested by graph (a), the climate sensitivity to a doubling of CO_2 is still only 1.3°C, and falls outside the IPCC's likely range (1.5°C – 4.5°C). Could the hockey stick curve be less convincing than it first appears?

Fortunately, exemplary *cherry picking* helps to restore conclusions which comply better with the IPCC's assumptions: by limiting the adjustment period to 1975–2000, a climate sensitivity of 3.1°C to a doubling of CO_2 can be obtained, and a warming rate of 0.2°C per decade (Figure 3.5).

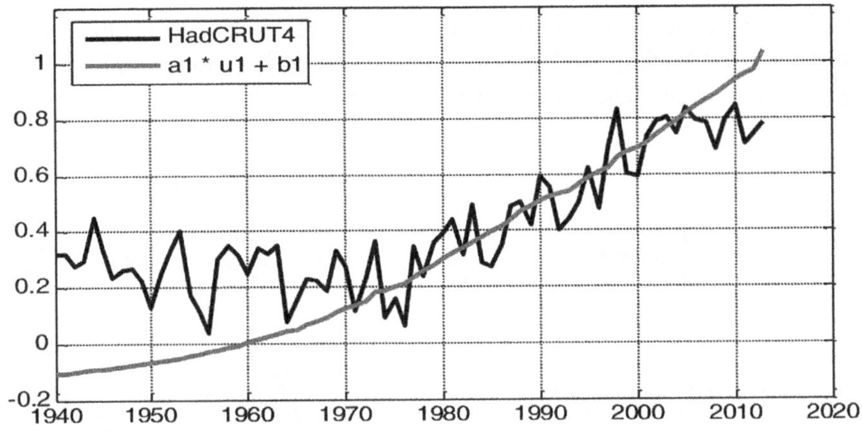

Figure 3.5. *Adjustment optimized over the period 1975–1998*

The decline before 1975, as well as the plateau emerging at the start of the 21st Century should however encourage a note of caution before setting alarm bells ringing.

Following the same train of thought, inconsistent conclusions can be drawn from Figure 3.4 (b). In particular, even at the scale of the figure, an eventual divergence can be seen between the two lines. This is even clearer in Figure 3.6 below, where Stauning [STA 11] corrects and expands a graph by Friis-Christiensen [FRI 91], previously limited to the period 1850–1985.

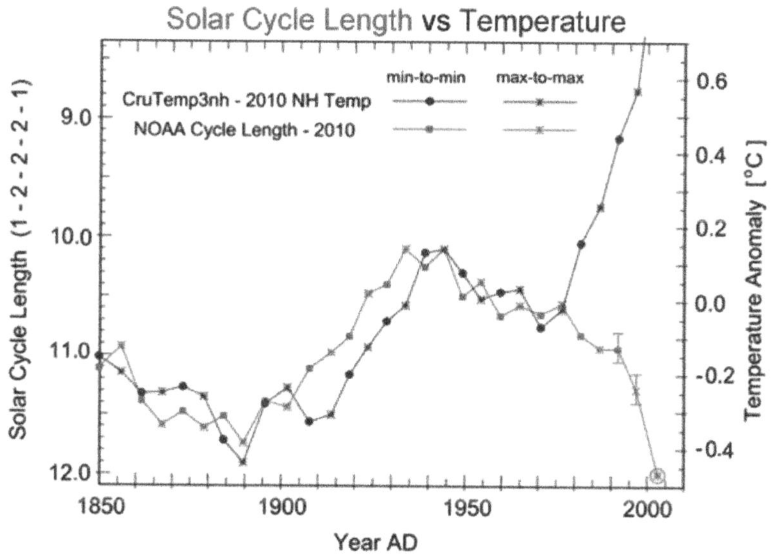

Figure 3.6. *Lines by [FRI 91], modified and expanded by [STA 11]*

This figure looks to deal a decisive blow against the theory of solar origins of warming, disproving the initial correlation with the final divergence[13]. In reality, the final gaps are scarcely greater than those which can be seen in Figure 3.4 (b) around the years 1000 or 1600, and do not challenge the correlation of large millennia-long climatic variations with solar irradiance.

Taking a closer look, the variations of the two lines 3.4 (b) are more often synchronized than not. The period 1980–2010 is simply in the part where the lines do not correlate. Looking even closer, (see Figure 7.4 below), a correlation appears between the Schwabe solar cycles (11 years) and temperatures.

In any case, the basis of all of these debates is inconsistent for a number of reasons. The first of which is that it is overly simplistic to polarize a theory towards a single cause: either "100% CO_2" or "100% solar".

13 The Frohlich and Lean correction (see Figure 2.10) in fact helps to accentuate the eventual divergence.

The second comes at the high level of random fluctuations, which can be seen in all the figures above. Observations which are limited to several decades – or even to one or two centuries – are not long enough to be able to use them for drawing definitive conclusions.

The third is that these overlays, regressions and correlations are only significant if the global climate machine is a process with no memory, where output depends only on the input at that precise moment, without taking past events into account. This is far from being the case, due to thermal inertia caused by the oceans. Therefore, even with the most sophisticated statistical techniques, the conclusions will be inconsistent, regardless of the theory being defended.

The only way of tackling the issue is by using a dynamical causal model. Indeed, correlations need to be established, not only between input and output at a given moment, but also between output and all or part of past input data. The technique for determining – explicitly or implicitly – such correlations is none other than that used for the identification of dynamical systems. Only this is able to establish whether, for example, the lines in Figure 3.5 contradict the theory of solar activity contribution or not. The signals still need to contain the necessary information, requiring that their duration be sufficiently long with regard to their spectrum; at least a millennium. It is also necessary to take into account all potential causes simultaneously, which multivariate identification automatically ensures.

3.3. Usable data

Despite its poor demonstration capability, Figure 3.4 may lead us to the worrying conclusion that the results of identification could differ greatly depending on the data used, and the battle of the graphs would give rise to questions over the overall validity of paleoclimatic data. It is not a question of challenging anybody's scientific rigor; when a fact appears to be perfectly established, it is admissible, in a pedagogical context, to select the observations which illustrate the argument in the clearest way, and even to emphasize the point. However, this is not justifiable when scientific fact is not yet established.

Finally, what credit should be given to the reconstructions available? In the AR5 (Figure 5.7), the IPCC presents the "plates of spaghetti", which are otherwise much more intricate than our Figure 3.2. They lead us to suspect, perhaps intentionally, that nothing but contradiction can be drawn from this collection of data. Nevertheless, the raw material (proxies, weather surveys or satellites) provides information which is more robust than it appears, and which cannot be so easily warped, even by the most adventurous statistical processing (as for the deliberate falsification of data, we needn't consider it: it is a practice reserved for politics). This is why we collected a range of measurements and reconstructions which, without being exhaustive, is representative of the entirety of the existing information on climatic history. If we are able to extract consistent results from it, the validity of our approach will be strengthened. And we will see that they rightly confirm, almost unanimously, the conclusions that we have put forward regarding the significant contribution of solar activity to recent warming. The only partial inconsistency element may come from the Jones and Mann reconstruction.

4

Formulating an Energy Balance Model

4.1. State models and transmittance

The first energy balance models appeared in the 1960s [ADE 64, BUD 69, SEL 69]. Typically, these models are based on the Stefan-Boltzmann law of radiation and on one-dimensional heat diffusion equations, according to latitude [GHI 76, NOR 75]. Other models, even more rudimentary, use single thermal inertia whose energy evolves depending on the planet's global energy balance. It is with this type of models – only a little more complex – that we will process here, and they are particularly suitable for identification.

The reader who is familiar with systems theory will excuse us for giving a brief presentation of the concepts of state and transmittance, using a suitable introductive example. We can imagine our planet reduced to its simplest form: a global thermal inertia (or heat capacity) I_G, brought to a homogenous temperature T_G by a balance of radiative flows Φ_R (index $_R$ for radiative) :

$\Phi_R = -\lambda_G\, x + \alpha\, u$, where:

– u is the variation of the solar constant in relation to its nominal value;

– αu is the flow of "radiative forcing" received from the sun, and α is a coefficient called radiative forcing, which depends on the coefficient of terrestrial reflection (albedo);

– $x = T_G - T_E$ is the deviation of the global temperature with regard to the equilibrium;

– $\lambda_G x$ is the infrared flow radiated, supposing that it is proportional to small temperature deviations.

The energy stored by the planet is the product $I_G . T_G$, and its variation is proportional to the balance of flows, hence $I_G \frac{dT_G}{dt} = \Phi_R$. The equilibrium temperature T_E, which being constant by definition, gives $\frac{d}{dt} T_E = 0$, and results from the equation for the model:

$$I_G \frac{d}{dt} x = -\lambda_G x + \alpha u$$

At any time t, we can consider that the variable $x(t)$ constitutes the *state* of this system, and the expression above is known as a *state space equation.* Here, it is a first order equation. The process defined thus is a linear time invariant system (LTI).

To an equation of this kind, we can associate a transmittance (or transfer function)[1]. This can be done by replacing the differential operator d / dt with the symbol s (the Laplace variable). The equation above then becomes $I_G s\, x = -\lambda_G x + \alpha\, u$, that can be processed like an ordinary algebraic equation. After factoring, $(I_G s + \lambda_G) x = \alpha\, u$, it finally becomes:

$$x = \left(\frac{\alpha}{I_G s + \lambda_G} \right) u$$

The expression $G(s) = \frac{\alpha}{I_G s + \lambda_G}$ is the transmittance associated to the initial equation of state. It is a rational fraction in s. It symbolically represents the process of transmission from a causal input u to the output x.

1 A presentation of these concepts, relevant to this work, is given in Chapter 3 entitled "LTI systems and transmittances" of the book *Automatique Appliquée* by Philippe de Larminat [DEL 09].

Here, it is a first order equation. By dividing the numerator and the denominator by λ_G, it can be written in the form:

$$G(s) = \frac{S}{Ts+1}$$

where $S = \alpha / \lambda_G$ is the static gain or equilibrium sensitivity of the output in relation to the input u_i, and where $T = I_G / \lambda_G$ is the time constant of the system. An equilibrium being a situation where all the derivatives are zero, $d/dt = 0$, the static gain of a transmittance $G(s)$ is given by $G(0)$.

A first order black box model can be defined by its transmittance $G(s) = \frac{S}{Ts+1}$, without any reference to "physical" coefficients, here α, λ_G and I_G. Parameters S and T characterize the step response of the process, that is, its response to a hypothetical variation in a step input of $u(t)$, as Figure 4.1 shows.

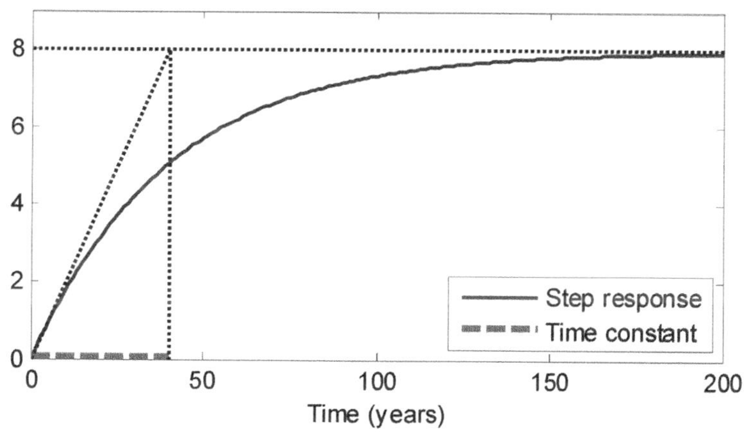

Figure 4.1. *Step response of a transmittance:* $G(s) = \frac{8}{40s+1}$

Observing only the external behavior of a system, it is impossible to determine the value of its physical parameters unless there is a minimal number of them. Otherwise, it would be necessary to arbitrarily fix one or the other, whereas we imposed on ourselves the rule of not using any *a priori* knowledge.

A complex system can result from the parallel association of a number of first order transmittances, or in series, or in feedback loops. The state becomes a vector ($x = \{x_1, x_2, \cdots, x_n\}$) Generalizing for nonlinear cases, the equations of state remain similar to what we presented: $dx/dt = f(x,\ u)$.

The global coupled atmosphere-ocean General Circulation Models (GCMs) result from physical laws of continuum mechanics (fluid mechanics and thermodynamics). They are described by a huge number of first order nonlinear state equations, resulting from their finite elements discretization. Their simulation requires gigantic calculation resources[2].

For the purposes of identification, we are looking instead for a minimal representation which is able to account for the input-output behavior of the process. This is the case for the more or less simplified energy balance models, such as the MAGGIC model [MEI 11], which has been extensively used by the IPCC since AR3, and also for the production of the RCPs (see 9.1). The number of finite elements of which they are made can be reduced to a few hundred, a few dozen, a few units, or even a single unit, like in our introduction model. The fundamental climatic concepts of radiative forcing, equilibrium sensitivity and climatic feedback, widely used by the IPCC, are inherent to this type of model.

4.2. Structure of an energy balance model

In a strictly black box context, the LTI model of a climatic process with inputs u_1, u_2, u_3 (log-CO_2, irradiance, volcanism) and output $x = T_G - T_E$ would be written in the more general form:

$$x = G_1(s)u_1 + G_2(s)u_2 + G_3(s)u_3$$

---

2 CNRM-CERFACS simulations require the use of 12 nodes (out of 32) of 8 processors of an NEC SX8 calculator from Météo-France working full-time for one and a half years, and a volume of data to process and store of around 400 To (400 thousand billion octets). IPSL simulations require the use of 3 nodes (out of 10) of 16 processors of an NEC SX9 calculator of the national apparatus for intensive calculations (GENCI) and a volume of data of 2 Po (2 million billion octets). A significant amount of national calculation power is therefore taken up by speculative climatic research, which is a different activity to short-term weather forecasting, also consuming large amount of calculation capacity.

where the three transmittances $G_i(s)$, and their respective static gains S_1, S_2, S_3, are free from any *a priori* constraints. We will recover a similar expression, using the generalization of our introductive example – which already contained the beginnings of EBM. Eventually, the structure above will be assigned with a restriction which seems minor, but is essential for ensuring the identification approach is carried out correctly.

Let us now consider the instantaneous balance Φ_R of the radiative energy flows entering and leaving at the TOA (Top-Of-Atmosphere). This makes the difference between the solar flow incident, and the flows reflected and re-emitted, both by the whole atmosphere and the surfaces of the ocean and land. All these flows are caused by climatic machinery, localized in the atmosphere and its immediate contact surface with the continents and oceans. The balance results from causes u_1, u_2, u_3, known as radiative forcing input, and from the global temperature anomaly $x = T_G - T_E$, then $\Phi_R = \Phi_R(u_1, u_2, u_3, x)$. Thanks to mixing caused by atmospheric circulation, it can be accepted that this function takes on value instantaneously (in less than one year).

The atmospheric mass also exchanges energy with continental surfaces (very little since the ground is an excellent thermal insulator), and especially with the enormous ocean mass. This is organized into different temperature and salinity layers, which float, circulate and slide under and over each other, very rarely mixing together. The state of the climate system can therefore not be reduced to the deviation x. The differences in temperature between the various layers compared to their own equilibrium temperature also needs to be included. As a result, the whole system is modeled using a system of differential equations based on the thermal inertias of the various layers, and the coefficients of resistance or thermal conduction which govern the energy flows exchanged according to temperature differences.

For a single thermal inertia I_G, we used $I_G \frac{d}{dt} x = \Phi_R$, or its equivalent transmittance formulation, $x = \frac{1}{sI_G} \Phi_R$. In the presence of a number of thermal inertias in a situation of linear interactions, an expression of the form $x = \frac{1}{sI_G(s)} \Phi_R$, would be used, where $I_G(s)$ is a complex thermal inertia whose coefficients combine the set of inertias and thermal exchange coefficients, by analogy with complex impedance $Z(s)$ of an electric network, function of elementary components R, L, C.

Reverting back now to the balance of flows Φ_R. However complex the function $\Phi_R(u_1, u_2, u_3, x)$ may be, it can always be developed in a linear approximation:

$$\Phi_R \sim \alpha_1 u_1 + \alpha_2 u_2 + \alpha_3 u_3 - \lambda_G x$$

– $\alpha_1, \alpha_2, \alpha_3$ are coefficients of radiative forcing;

– the products $\alpha_1 u_1, \alpha_2 u_2, \alpha_3 u_3$ are flows of radiative forcing;

– the sum $Q = \alpha_1 u_1 + \alpha_2 u_2 + \alpha_3 u_3$ is the total radiative forcing;

– the term $-\lambda_G x$ is a global climatic feedback.

The coefficient λ_G gives the sum of various elementary reactions. Some are stabilizing, in counter-reaction and have a positive sign in λ_G. Destabilizing reactions act with a negative sign: an initial deviation on x leads to an increase of Φ_R, which in turn causes an increase of x, and so on, hence the divergence.

Eliminating Φ_R, the equations are combined:

$$I_G(s)\, s x = \alpha_1 u_1 + \alpha_2 u_2 + \alpha_3 u_3 - \lambda_G x$$

Hence:

$$x = \frac{1}{\lambda_G + s I_G(s)} (\alpha_1 u_1 + \alpha_2 u_2 + \alpha_3 u_3)$$

Setting down $S_1 = \frac{\alpha_1}{\lambda_G}$, $S_2 = \frac{\alpha_2}{\lambda_G}$, $S_3 = \frac{\alpha_3}{\lambda_G}$ and $G(s) = \frac{\lambda_G}{\lambda_G + sI(s)}$, finally gives:

$$x = G(s)(S_1 u_1 + S_2 u_2 + S_3 u_3)$$

where $G(s)$ here is a transmittance of a unit of static gain: $G(0) = 1$.

4.3. Specificity of EBMs

The most general form given for a black box model was $x = G_1(s)u_1 + G_2(s)u_2 + G_3(s)u_3$. In the case of an EBM, a transmittance $G(s)$ is found as a common factor. It is not only a case of a reduction of the same denominator of the $G_i(s)$, which would leave different numerators. The result is a reduction in complexity, and therefore in the number of parameters to identify. Of course, this reduction is a reaction to the practical reality of the climatic process, and not to an abstract theoretical imperative. In fact, it is inherent to the concept of the energy balance model itself.

The common transmittance $G(s)$ means that the global temperature responds with the same transients to different sources of radiative forcing. The global temperature would evolve in the same way to any supposedly identical variations, no faster or slower, as to a volcanic event or a variation in CO_2. This is the case because radiative forcing always develops at the same place: the atmosphere and the thin layer of continental and ocean surface. There is no direct radiative flow between deeper layers and space. All components of the global radiative balance $\Phi_R \sim \alpha_1 u_1 + \alpha_2 u_2 + \alpha_3 u_3 - \lambda_G x$ are therefore transmitted in the same way to the deep ocean layers, whatever the nature of the forcing u_i. The different forms of radiative forcing u_i thus only have an effect on static gains, and not on the transients of the response.

Let us now return to the global climatic feedback coefficient λ_G, which is common to all the forcing inputs. Remember that this coefficient is the sum of various contributions: stabilizing reactions contribute with a positive sign in λ_G, destabilizing reactions with a negative sign. The most important of these reactions results quite simply from the infrared radiation emitted by the terrestrial surface: the more positive the deviation $x = T_G - T_E$ is, the higher the infrared radiation emitted is. The greenhouse effect may dilute this effect, but could certainly not reverse it.

A destabilizing effect often put forward is that of the greenhouse effect by water vapor, where a higher temperature increases the partial pressure of H_2O (greenhouse gas) in the atmosphere. However, at the same time, cloud cover increases and therefore so does the reflection of solar rays, leading to a simultaneous stabilizing effect. It is difficult to physically determine which of the two dominates. There must be a wide margin of stability due to primary radiation increasing with temperature, since the planet has not been definitively transformed into a glacier or a boiler under the influence of volcanism or variations of CO_2 that it has experienced throughout the Earth's history.

The terrestrial albedo also contributes in several fundamentally different ways. Its average "nominal" value is 30%. It may vary from this for a number of reasons: 1. due to the effect of deforestation which is an independent anthropogenic cause. The result is not a reaction, but a radiative forcing (and the changing use of land should also be added to the list of causal inputs outlined); 2. due to the effects of temperature variation T_G itself, including the lower increase for snow-covered tundra and Canadian plains. Hence this time it is a destabilizing reaction, giving a negative contribution in the coefficient λ_G. The albedo brings solar irradiance into play, but the reaction above is nonetheless related to the variation in global temperature. The variation in irradiance only has a second-order effect, which by definition is not taken into account in the linear approximation of the total TOA flow.

These subtleties are not easy to grasp. To summarize: there is no specific feedback for a given input of radiative forcing, whether for CO_2, irradiance or volcanism; the temperature T_G remains the intermediary from which the reaction is taken, whatever the primary cause of its variation. Confusion often arises – and not only among non-specialists. As a result, many people, starting from an increase in CO_2, methodically follow step by step the effects on temperature T_G, the increase in partial pressure of H_2O which results from this, and so on (while usually relying on destabilizing reactions). However, in any case, the starting point is irrelevant: once a variation in T_G is obtained, the rest of the causality chain remains identical, whether the source is CO_2, variations in volcanic activity, or random disturbances of atmospheric or ocean currents. Supposing that the climatic machine does not have a wide margin of stability, the impact of CO_2 is not necessarily more destabilizing than any other.

The terms of change themselves, especially for climate disruption, damage the concept behind the model itself: they imply that some parameters, in particular λ_G, may be time-dependent[3], which serves no purpose if we do not specify what they vary in relation to. This is how [BOE 05] consider that climate sensitivity is dependent on the level of forcing. Such a dependence in relation to the variables u_1, u_2, u_3 or T_G therefore implies nonlinearity, to be processed as such, or to be integrated in an LTI approximation.

Ultimately, the coefficient λ_G is what it is. Excepting incoherence, it is constant. There are as many reasons to enlarge it as there are to reduce it. It is, above all, very uncertain, whether globalized in a reduced EBM, or distributed through a large-scale GCM.

In the context of identification, the coefficients $\alpha_1, \alpha_2, \alpha_3$ and λ_G will unfortunately be individually inaccessible using the observation of input/output data. Only the transmittance $G(s)$ can be identified along with sensitivities in the form of ratios $S_1 = \dfrac{\alpha_1}{\lambda_G}$, $S_2 = \dfrac{\alpha_2}{\lambda_G}$, $S_3 = \dfrac{\alpha_3}{\lambda_G}$. This does not prevent us from taking advantage of the fact that static sensitivities S_i are in the same ratio as radiative forcing coefficients α_i, but to increase to three α_i and λ_G, it would be necessary to have prior knowledge of one or other of the four parameters. Unfortunately, as will be seen in Chapter 5, there is not one of these which physical knowledge allows us to determine *a priori* with enough accuracy.

4.4. Dynamic parametrization

The above formulation of equilibrium sensitivities can be found, at one level or another, in all climatic energy balance models. It is also implicitly

3 AR4, end of paragraph 8-8-2: The obtained best-fit climate sensitivity estimates differ for various reasons from other estimates that were derived with alternative methods. Such alternative methods include, for example, regression estimates that use a global energy balance equation around the year of atmospheric CO_2 doubling or the analysis of slab ocean equilibrium warmings. The resulting differences in climate sensitivity estimates can be partially explained by *underscore* in many of the AOGCM runs. Furthermore, tuning results of a simple climate model will be affected by the model structure, although simple and other default parameter settings that affect the simple model transient response.

present in GCMs, insofar as they model a linearizable climatic process as a first approximation.

However, the form of the transmittance $G(s)$ is specific to each type of EBM. It depends on its structure; the definition of finite macro-elements adopted: continents, oceans, hemispheres, cryosphere, atmospheric or ocean layers, and so on. It combines, in a complex way, thermal input and coefficients of exchange between these compartments. Insofar as EBMs, like GCMs, appear to be representative of the same climatic reality, all simulated responses will necessarily have the same appearance.

As a result, the following curve (Figure 4.2) is taken from *Stowasser* (2006). It comes from the simulation of a GCM[4] driven by a step of solar irradiance of 5%, which corresponds to a net variation of $u_2 = 12\ Wm^{-2}$. The final temperature is 8°C, (possibly more, if the text is to be believed). The coefficient of climate sensitivity to net irradiance $S_2 = \Delta T / u_2$ would therefore be equal to $0.666°C / Wm^{-2}$, a value which complies with the IPCC's estimated values (see section 5.5 below).

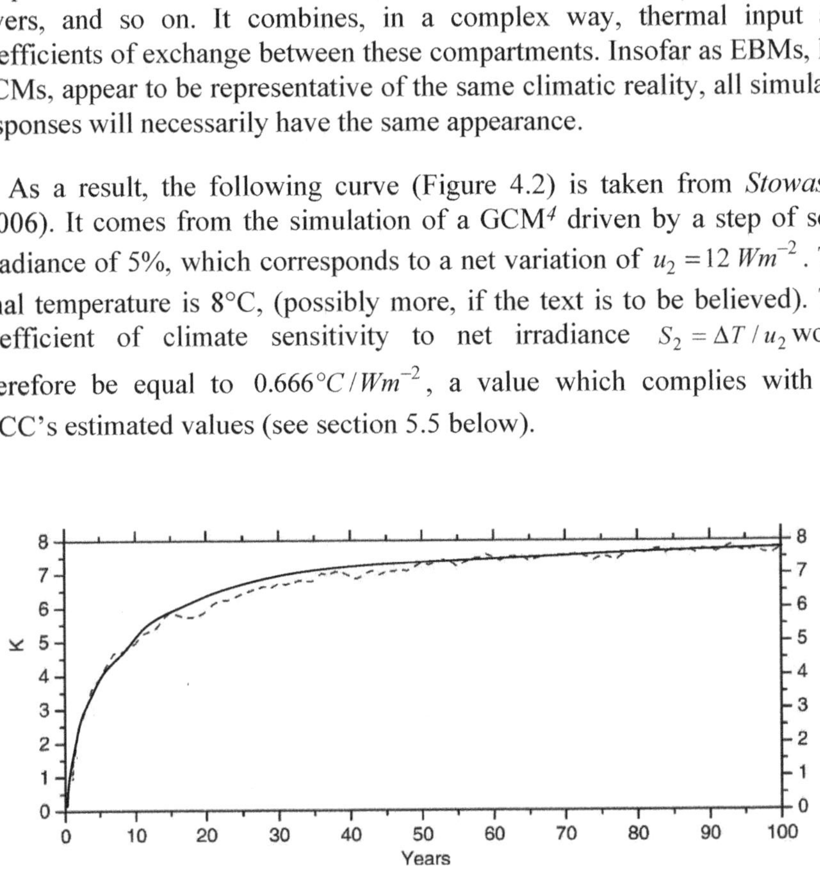

Figure 4.2. *Climatic response simulated by GCM (---) and rational approximation (—)*

4 Coupled General Circulation Model version 3 of the Canadian Centre for Climate Modelling and Analysis (CCCma)

If the final value is reduced to the unit, the response above defines the step response (the response at a unit level) of the transmittance $G(s)$. Having arrived here via a GCM, its order is *a priori* very high. Fluctuations resulting from the appearance of chaos are inherent to natural turbulent flows and their numerical simulations. The underlying response obeys the general behavior of heat propagation: very fast initial rise, and ultimate transitory phases which return to their asymptote extremely slowly, typical of thermal exchange processes.

Nevertheless, its modeling with an idealized (fractal) equation of heat diffusion is not required (see Van Hateren, [VAN 13]), since it is no more compliant than any other to the complex physical reality of the climatic machine.

Ordinary tools for identification usually allow transmittances in the form of a ratio of any two polynomials in s, which gives too much latitude to the identified models: instabilities, resonant modes or non-minimum phases. Moreover, optimizations required by identification techniques often appear to be poorly conditioned in relation to denominator coefficients. We therefore suggest defining $G(s)$ as the sum of *a priori* fixed time constants T_i, where only the positive multiplicative coefficients c_i are free:

$$G(s) = c_1 \frac{1}{1 + s\,T_1} + c_2 \frac{1}{1 + s\,T_2} + \cdots + c_n \frac{1}{1 + s\,T_n}$$

Thanks to a sufficient amount of time constants T_i, it is possible to approximate any aperiodic system. In practice, three or four time constants in geometric progression are enough to represent a system with multiple scale systems. Therefore, the approximation given on Figure 4.2 corresponds to the transmittance response $8.3 \times G(s)$, where:

$$G(s) = \frac{0.2}{1+s} + \frac{0.6}{1+10s} + \frac{0.2}{1+100s} \quad \text{(where } c_1 + c_2 + c_3 = 1\text{)}$$

The climate would be likely to keep a memory of its past through even greater time constants, of around a millennium or more, and we had planned to complete the series by a term in $1/(1+1000s)$. However, it then becomes practically impossible to distinguish the unknown equilibrium temperature T_E from the memory effect associated to this time constant. Here, we prefer to keep T_E as a parameter to estimate, and limit the series to [1, 10, 100 years]. Finally, the model is therefore written:

$$T_G = T_E + \left(\frac{c_1}{1+s} + \frac{c_2}{1+10s} + \frac{1-c_1-c_2}{1+100s} \right)(S_1u_1 + S_2u_2 + S_3u_3)$$

Of course, we checked the robustness of the identification with respect to the number and the choice of the T_is. The number of parameters to identify (six, counting T_E) appears to be sufficient for modeling the fundamentals of the climatic process.

Conversely, we could have attempted to distinguish the terrestrial and ocean temperatures, or those in the northern and southern hemispheres. However, firstly, the individual temperature databases only exist from 1850, and secondly, the ratio of the number of new measurements to the number of new parameters introduced would lead to an over-parameterization, which is prohibitive in the context of black box identification. Expect for some minor variations, regarding the parametrization of $G(s)$, the model adopted here seems to be practically indispensable (and Van Hateren's is very close to it).

Dynamic parameters

The IPCC has introduced a parameter named TCR: Transient Climate Response. It is defined as the rise in temperature obtained after 70 years, in a scenario of annual CO_2 increase of 1%. These 70 years correspond to the duration necessary for a doubling in the concentration of CO_2 $(1.01^{70} \sim 2)$. To date, the equilibrium having not yet been reached, we are thus obviously left with $TCR < S_{clim}$.

The TCR parameter is directly proportional to S_{clim}, and the TCR/S_{clim} ratio is a complex function of all the temporal and frequential features of transmittance $G(s)$. The smaller it is, the slower the climatic process is to react to the influence of forcings, whatever they may be: CO_2, TSI, or AOD. Using this ratio alone, it is impossible to include all features that are of interest to us.

Nevertheless, the collection of all the characteristics of $G(s)$ must be reduced to the most pertinent. This is why we prefer to associate the following two dynamic features to the transmittance $G(s)$:

– A global time constant T_{clim}, defined as the final delay (or lag) for the response to a ramp input:

$$T_{c\lim} = \lim_{s \to 0}\left(\frac{1-G(s)}{S}\right)$$

– The ratio ρ of high and low frequency sensitivities: more precisely, the "high frequency" considered will be that of the Schwabe cycles (f = 1/11 years), which is effectively high at the time scale of the climate. This is expressed as:

$$\rho = \left|G(j2\pi/11)\right|/G(0)$$

We will use this coefficient particularly for solar irradiance. If we note S_{irr11} the sensitivity to the frequency $f = 1/11$, this will be given by $S_{irr11} = \rho S_2$. This ratio will be small (around 0.1 or less), but not negligible.

5

Presumed Parameters

The aim of this chapter is to draw out, through the major IPCC documents, in particular AR5 and SPM, the climatic parameters' values as well as their ranges of uncertainty, with a view to comparing them to the results of identification.

5.1. Terminology

In the IPCC documents, certain qualifying terms are systematically given in italics: *likely, very likely*, or even: *high confidence, low confidence*, and so on. They refer to a precise and restrictive scale (Mastrandrea *et al.*, 2010). They may apply to the quantified value of a variable, or to any kind of assertion.

Regarding a variable, quantified by an equality or inequalities, the probabilities are expressed using the following scale:

Virtually certain	99–100 %
Extremely likely	95–100 %
Very likely	90–100 %
Likely	66–100 %
More or less equally likely	33–66 %
Unlikely	0–33 %
Very unlikely	0–10 %
Extremely unlikely	0–5 %
Exceptionally unlikely	0–1 %

When the IPCC specifies a confidence interval, it is generally 90% (*very likely*)[1]. The sum of the probabilities to exceed the given upper or lower limits is therefore 10%, without them necessarily being equally shared at 5% each. It should not be assumed either that the most likely value is situated in the middle of the range.

Regarding an assertion (not quantified), the degree of confidence results from both the reliability of the indicators (type, quantity, quality and consistency), codified as *limited, average* or *robust*, and from the degree of agreement amongst the experts in their collective judgment: *weak, average, high*. Hence, skillfully combining these two factors, gives a scale of overall confidence:

Very low	10 %
Low	20 %
Medium	50 %[2]
High	80 %
Very high	90 %

Since giving a level of confidence is an assertion, it is generally subject itself to an assessment of confidence (see section 5.2), to which it in turn could be subject...but let's stop there.

The probabilities indicated (in %) may result from effective statistical calculations, especially the ranges of confidence, but not necessarily. Conversely, it is clear that a level of agreement between the members of a group of experts has nothing to do with a formal statistical calculation based on observed data, even though providing evaluations in % in the context of a scientific report may strongly suggest that it results from statistical calculations.

Moreover, some climatic parameters arise from a combination of other parameters (e.g.: $\lambda_G = \log(2)\alpha_1 / S_{\text{clim}}$). In most cases, even using fuzzy logic, it will be impossible to accurately combine the evaluations of the

1 Statisticians often prefer 95% (for confidence ranges of +/- two standard deviations in the Gaussian case).

2 When the IPCC gives the label of *medium confidence* to some catastrophe caused by climate change, this allows them to announce it without taking the risk of an explicit assertion.

assumptions. The final evaluations will thus only carry over an informal degree of confidence for a given evaluation.

5.2. Climate sensitivity S_{clim}

Looking once again at the SPM: "Equilibrium climate sensitivity is the change in global equilibrium temperature caused by a doubling in the concentration of CO_2 in the atmosphere. It is likely to be in the range of 1.5°C to 4.5°C, (*high confidence*), extremely unlikely to be under 1°C, (*high confidence*), and very unlikely to be above 6°C (*medium confidence*)".

There is nothing to indicate that the confidence level of *likely* (at 66%) would be symmetrical, especially given that the upper and lower limits are unsymmetrical (respectively 95-100% and 90-100%), and are weighted respectively to either *high* or *medium* confidence levels.

The AR4 (2007) gave 3°C as the most likely value, which was not centered in the middle of the given range [2 to 4.5°C]. The second order draft of the AR5 (Chapter 12, p 7, line 18) again stated that, "the most likely value remains near 3°C". Due to a lack of consensus, this was removed from the final version of the AR5. It would therefore be unsafe to infer from all of the information above that there is a hypothetical value which equates to the *most likely*. We are clearly in a situation where probabilities which are given do not arise from actual statistical calculations, but from negotiations between experts.

If we only retain the extreme values:

$$1\,°C < S_{clim} < 6\,°C\,,$$

after dividing by log(2), it follows:

$$1.4427\,°C < S_1 < 8.6562\,°C$$

When comparing these ranges to our own confidence level (at 90%), we will bear in mind that those of the IPCC are hindered by a certain level of fuzziness.

5.3. Coefficient of radiative forcing α_1

The AR5, section 8.3.2.1, states: "Using the formula from Table 3 of Myhre *et al.* (1998) [...], the CO_2 Radiative Forcing from 1750 to 2011 is 1.82 (1.63 to 2.01) Wm^{-2}. The uncertainty is dominated by the radiative transfer modelling which is assessed to be 10%". Curiously, these assertions are not assigned a level of confidence – that we personally feel should be positioned at the lowest level (see section 5.8).

Table 3. Simplified expressions used in *IPCC* [1990] (Table 2.2)

Trace gas	Simplied expression Radiative forcing, ΔF, Wm^{-2}	Constants α IPCC	Best estimate this work[a]
CO_2	$\Delta F = \alpha \ln(C/C_0)$	6.3	5.35

Figure 5.1. *Taken from Myrhe (1998)*

This table (Figure 5.1) confirms that forcing by CO_2 follows a logarithmic law, and specifies the value of the coefficient of radiative forcing:

$$\alpha_1 = 5.35\ Wm^{-2}$$

For a doubling of CO_2, by multiplying α_1 by log(2), we obtain a radiative forcing of 3.7 Wm^{-2}. This value is taken up in AR3, AR4 and AR5, and has not changed since 1998.

5.4. The climate feedback coefficient λ_G

The climate feedback coefficient λ_G affects the denominator of expressions $S_i = \alpha_i / \lambda_G$. Hence the interest in investigating its inverse: *global sensitivity to radiative forcing*:

$$S_G = \lambda_G^{-1}$$

Given that $S_G = S_1 / \alpha_1$, it would be necessary to establish the joint probability of S_1 and α_1 in order to obtain the range of uncertainty of the quotient. But since the relative uncertainties relating to S_l ($1.4427\,°C < S_1 < 8.6562\,°C$) are much greater than those for α_1 ($\pm 10\%$), it allows us to simply divide the limits of S_1 by α_1:

$$0.27°C / Wm^{-2} < S_G < 1.62°C / Wm^{-2}$$

Reversing and reordering gives the limits on λ_G:

$$0.62\ Wm^{-2} / °C < \lambda_G < 3.7\ Wm^{-2} / °C$$

We are particularly interested in the upper limit of S_G. By referring to the initial statement on climate sensitivity, we can write:

$$S_G < 1.62°C / Wm^{-2}\ (likelihood\ 90 - 100\ \%).$$

5.5. Sensitivity to irradiance S$_2$

Since irradiance is evaluated in Wm^{-2} *net*, the forcing coefficient for solar activity α_2, restricted to the energy factor only, is by definition equal to 1. Given that $S_2 = \lambda_G^{-1}\, \alpha_2$, the range for sensitivity to net irradiance is therefore identical to that of S_G:

$$0.27°C / Wm^{-2} < S_2 < 1.62°C / Wm^{-2}$$

We are particularly interested in the upper limit and we can therefore state, in compliance with IPCC terminology:

It is *very unlikely* that sensitivity to net irradiance is higher than $1.62\ °C / Wm^{-2}$.

The IPCC did not explicitly make this assertion, but it can be inferred from the ranges indicated for S_1 and α_1, reviewed, corrected and confirmed by successive IPCC reports, as well as the ratios of $S_i = \lambda_G^{-1}\, \alpha_i$, which themselves arise from the fundamental concepts of the IPCC: radiative forcings, energy balance and global climate feedback.

The evaluation itself is in accordance with the IPCC's constant claim that, since 1950, natural factors have only had a minimal effect of global warming[3]. Since the variations in net irradiance are otherwise very low, just fractions of Wm^{-2}, the influence on the climate would be a few hundredths of a degree, while the impact of CO_2 can be counted in at least tenths of a degree.

Total sensitivity to solar activity

In the draft SPM, a version of which was leaked before its release[4] (p. 8, line 41), a detailed account of climate change drivers stated that "cosmic rays enhance aerosol nucleation and cloud condensation nuclei production in the free troposphere, but there is high confidence that the effect is too weak to have any significant climatic influence".

Nevertheless, the same text adds that: "Many empirical relationships have been reported between GCR or cosmogenic isotope archives and some aspects of the climate system [...] but the forcing from changes in total solar irradiance alone does not seem to account for these observations, implying the existence of an amplifying mechanism such as the hypothesized GCR[5]-cloud link". However, it later (Chapter 10) states "high confidence that the effect of changes in solar irradiance are much smaller than the climatic changes resulting from a rise in radiative forcing due to GHGs".

The expression "amplifying mechanism" gives rise to confusion: it suggests that an amplification could act specifically on the solar energy flow, whereas "amplifying mechanisms" are common to all radiative forcing, through the common reaction coefficient λ_G, the sum of all climate reactions, whether positive or negative (see section 4.2). The "hypothetical link" between the GCRs and clouds would therefore not be a mechanism which *amplifies* the irradiance, but an *alternative* mechanism of solar action. The link to GCRs is mentioned several times in the AR5, without ever alluding to the CLOUD experiment (Kirkby, 2011, Almeida, 2013), carried out at *CERN* from 2009 to 2011 by around fifteen different institutes. Finally, the reference above to observations, correlations and GCRs was deemed inconvenient by the IPCC, and was removed from the final version

3 See "*Change in solar irradiance*", figure *SMP*.5.

4 http://www.stopgreensuicide.com/SummaryForPolicymakers_WG1AR5-SPM_FOD_Final.pdf.

5 Galactic Cosmic Rays, which we know are affected themselves by solar magnetism.

of the SPM: in the exhaustive list of radiative forcing (pages 11 and 12), no reference is made to any solar forcing other than energy flow itself. Everywhere else, a *high level of confidence* is reaffirmed regarding the fact that variations in solar irradiance could only have a minor influence on climate change.

To our knowledge, none of the IPCC models (GCM or EBM) include an "amplifying mechanism", nor any other mechanism, whether physically based or empirical. In conclusion, the sensitivity to solar activity, for the IPCC, is therefore reduced to only the sensitivity to energy flow. The upper limit defined above ($1.62°C / Wm^{-2}$) thus quantifies the IPCC's official position.

On the contrary, since our goal is to compare these assumptions to observations, we will not impose an upper limit for sensitivity to irradiance in our algorithms, even if the result may be identified values which significantly exceed this assumed limit.

5.6. Sensitivity to volcanic activity S_3

Nowhere in the IPCC reports is the coefficient of radiative forcing quantified, nor is the sensitivity to volcanic activity (measured by optical depth). However, a number of authors, including Hansen (2002), have looked into the issue. Wigley, *et al*., (2005), cited by Gao (2008), give the successive values 30 Wm^{-2}, 27 Wm^{-2}, 21 Wm^{-2}, before settling on 20 Wm^{-2} (per unit of optical depth) as a "somewhat arbitrary" estimate for the conversion factor α_3 of the optical depth in radiative forcing.

Multiplying the bounds $0.27°C / Wm^{-2} < S_G < 1.62°C / Wm^{-2}$ by 20 Wm^{-2} gives the range for S_3:

$$5.4°C < S_3 < 32.4°C$$

Identification will confirm this order of magnitude.

5.7. Climate or anthropogenic sensitivity

The concentration of CO_2 is the only input in our model which is related to human activity. We know that this also entails the emission of other greenhouse gases (methane, etc.), by industrial aerosols, or by changes in land use, the impact of which then adds or subtracts from the influence solely restricted to CO_2.

Adding these elements as distinct inputs into an identification model would cause a number of problems. Firstly, the various greenhouse gases are not driven by identical laws (logarithmic for CO_2, square root for methane, etc.). Moreover, some factors have an impact due to their concentration, others by their emission rate, which effects whether the time they remain in the atmosphere is respectively short or long. The concept of GWP (Global-Warming Potential) attempts to bring together their effects. In any case, they are all highly correlated between each other, due to the simple fact that, since the start of the industrial era, they have all grown exponentially. By identification, it would therefore be difficult to distinguish between their respective effects.

If we limit anthropogenic inputs solely to the concentration of CO_2, this must then be interpreted as an indicator of human activity, and the sensitivity identified will aggregate the sensitivity to all the other factors not listed, including all the approximations stated above.

The identified sensitivity S_{clim} will then be an anthropogenic sensitivity, whose assumed value it is important to establish. For this reason, we have examined the information provided by the IPCC on the various anthropogenic forcings listed. Amongst them, some are positive (other GHGs), others negative (aerosols and variations in albedo due to a change in land use). The overall result is likely to be positive, which would move the range upwards. Given the uncertainties, the extreme lower limit is maintained ($1\,°C < S_{clim}$). Since there is only *medium confidence* in the upper limit, this is also maintained ($S_{clim} < 6\,°C$), rather than resorting to risky calculations.

In the end, no distinction will be made between the ranges of uncertainty for climate sensitivity in the strictest sense (relating to CO_2) and for anthropogenic sensitivity.

5.8. Review of uncertainties

As stated earlier, the assumed values, attributed here to the IPCC, are based on its own quantifications, and its own fundamental concepts of radiative forcing, static sensitivity and climatic feedback. Nevertheless, the IPCC implies that the uncertainties could be significantly greater than stated.

Indeed, following discussions during the meeting for the SPM[6], the Coordinating Lead Authors (CLA) complements the definition of climate sensitivity as follows:

"The lower temperature limit of the assessed likely range is thus less than the 2°C in the AR4 (2007), but the upper limit is the same*. The assessment reflects improved understanding, the extended temperature record in the atmosphere and ocean, and new estimates of radiative forcing". * In a footnote: "it is not yet possible to say which is the best estimate for climate sensitivity, due to a lack of agreement between values resulting from various studies and demonstration proposals."

In the fourth SPM (2007), the most likely value was mentioned (3°C). Rather than revise it downwards, it disappeared from the SPM 2013.

New estimates for radiative forcing are mentioned. However, since AR3 (2001), the IPCC has used the same formula (Myrhe, 1998), without changing even one decimal, as if it was a definitively established physical constant. Although the IPCC reports on a significant amount of later work, a quantified consensus cannot be established. In the AR5, it is stated:

8.3: "The concentrations of WMGHGs can be determined from observations [...]. Thus the radiative forcings from WMGHGs are determined entirely from observations."

---

6 On equilibrium climate sensitivity, several delegations, including Australia, the Netherlands and others, noted that the message that the lower limit of the assessed "likely" range of climate sensitivity is less than the 2°C in the AR4 can be confusing to policy makers and suggested noting it as the same as in previous assessments. The CLAs explained that comparison to each of the previous IPCC assessments would be difficult, and new language was developed adding that the upper limit of the assessed range is the same as in AR4. (Kosolapova, 2013). This teaches us that values given for fundamental climatic parameters for the planet are negotiated during the plenary session of the IPCC, between scientists and governmental delegations.

Entirely? Whilst the laws adopted to describe the greenhouse effect come but from calculations of physical theory. Moreover, these have been contested (Miskolczi, 2007) and are above all unrealistic, as admitted by the authors of the report themselves:

8.3.1: "Most intercomparison studies of the RF of GHGs are for clear-sky and aerosol-free conditions; the introduction of clouds would greatly complicate the targets of research and are usually omitted in the intercomparison exercises [..]. It is shown that clouds can reduce the magnitude of RF due to GHGs by about 25%".

Where this value of 25% comes from is not specified, nor is the reason why this reduction could not be significantly higher (why not 75% or any other value?). Indeed, they add: "We underscore that uncertainty in RF calculations in many GCMs is substantially higher owing both to radiative transfer codes and meteorological data such as clouds adopted in the simulations."

It is also worth looking closely at the logarithmic law, assumed to apply up to 10,000 ppm (1% of CO_2). Complete saturation could occur well before, if it has not already been reached.

Uncertainties regarding sensitivity $S_1 = \alpha_1 / \lambda_G$ do not result only from the numerator α_1, but also due to the lack of robustness regarding the climatic reaction coefficient λ_G. Remember that this has an effect on the linear approximation of the TOA flow, and is the sum of various terms. Clouds play an essential role in this issue and according to the IPCC, "there is a *very high confidence* that the uncertainties in cloud processes explain much of the spread in modeled climate sensitivity" (AR5, Chapter 9, p.743).

The effect of the inverse of λ_G on climate sensitivity, $S_G = \lambda_G^{-1}$, explains the lack of symmetry in the uncertainties. Obviously, the closer the lower limit of λ_G is to zero – while remaining positive – the greater the relative uncertainty is for its inverse. Hence the *medium confidence*, and nothing more than this, given by the IPCC to the upper limit of climate sensitivity. Conversely, the recent reduction by the IPCC of the lower limit of climate sensitivity means that the stabilizing effects had previously been underestimated. Ultimately, the feedback coefficient λ_G would not only be very uncertain, but, according to R. Lindzen [LIN 01], significantly underestimated by the IPCC.

The uncertainties relating to λ_G also influence sensitivity to irradiance $S_2 = \alpha_2 / \lambda_G$, where α_2 is postulated to be equal to 1. Yet, we know (Svensmark, [SVE 00]) that other mechanisms could exist through which solar activity may have an effect on the climate. These mechanisms have not yet been well highlighted or quantified [KIR 11, ALM 13]. The IPCC can therefore not include them in its GCMs. However, an identified EBM is not subject to such constraints.

In conclusion, we have decided to wipe the slate clean of all *a priori* evaluations, and to identify climate parameters and their confidence ranges without imposing any constraints. Afterwards, we will compare them with the assumed evaluations, and in particular with the upper limit put forward relating to sensitivity to irradiance.

6

Identification Method

6.1. The current state of affairs

Not all GCM parameters came from fundamental physics, and some need to be adjusted, more or less empirically, like those relating to the genesis of clouds and their feedback effects, or the biochemical processes of vegetation, or micro-turbulence, or many others. Centers of modeling give little information on the way in which they adjusted the parameters of their model. Obtaining globally stable and realistic temperature levels in simulation is in itself not a trivial matter. Large-scale identification of these models is out of reach[1]. However, identification limited to a small number of critical parameters, known as closure parameters, can be envisaged [JAR 10]. This can be done by using techniques of identification, for instance by nonlinear filtering (UKF : Unscented Kalman Filter). Nevertheless, we are still a long way from a black box type identification, and the result is also highly dependent on the quality of the rest of the model. An even more restricted approach consists of detecting and attributing to a given input the responsibility in a given element of observed climatic behavior: typically, to detect the "fingerprint" of various human activities in warming.

Complete black box type identification can only be achieved with models of reduced complexity, like those presented in Chapter 4, and it is reasonable

1 According to the author of *box 9.1* of the AR5, "the need for model tuning may increase model uncertainty. There have been recent efforts to develop systematic parameter optimization methods, but owing to model complexity they cannot yet be applied to fully coupled climate models".

to expect that numerous attempts have already been made since the appearance of paleoclimatic reconstructions. In fact, the very first EBMs were not carried out for the purposes of identification, but rather to perform abstract theoretical analysis. They very quickly became mini-models of general circulation, with large numbers of parameters and therefore unidentifiable. For instance, when Crowley [CRO 00] compares the MAGGIC model to paleoclimatic data, he adopts the parameters assumed by the IPCC. Even though some parameters are subject to variations designed to study their effect on the models' behavior, this is not true identification.

However, there is one case where "official" techniques of identification (Ljung, 1999) are applied to EBMs, but in a context of simplifying complex models by identification [WEL 14]. This is carried out, not by using real historical data, but with data obtained by simulation of GCMs. The goal is therefore to obtain more flexible models, whose global behavior is identical to that seen in GCMs, and also to extract from them values for global climate parameters; the first being the fundamental parameter: climate sensitivity.

And yet, why not carry out identification using authentic historical data? Firstly, because they are less rich in terms of information compared to simulated signals on which one can "experiment" repeated variations for forcing input, and also possibly because the results do not comply with the intents: they do not clearly confirm the anthropogenic influence, or they show the predominant impact of solar activity. In summary, the AR5 makes no reference to any identification of a climate model, whether black box or grey type, neither in Chapter 9, regarding the evaluation of models (EBM or GCM) nor elsewhere.

Nevertheless, we are taking a leap into the unknown. The model – or class of models – that we propose to identify is intended to be completely free of any *a priori* knowledge, except the general structure of the model (Chapter 4). Of course, we will make the connection between the parameters of the model and the climatic parameters familiar to all climatologists. We will also occasionally make the link between some features of the model and general behavior of thermal processes, or even to propose related explanations, for instance, for thermohaline circulation. However, in no case will the identified models rely on the accuracy of our knowledge of climatic processes, which is an area in which we do not claim to have any expertise.

The models that we are able to identify will come from data, and solely from climatic data.

6.2. Output error method

Following the data collection process, the identification of a process could be reduced to a style exercise in which we determine, according to the rules of the art, a causal mathematical relation between the input and output signals. Nevertheless, since climatic databases are significantly noised and relatively sparse in terms of events, it would be unrealistic to attempt to identify a model with more than 6 or 7 parameters. This would systematically exclude so-called non-parametric techniques, which identify point by point frequential or temporal responses.

The most straightforward parametric identification method is that of least-squares output error. This comes to mind naturally, without needing to make reference to any pre-existing theory: it solely consists of minimizing, with respect to the parameters of a model, a norm of deviations between the output observed and the input simulated by the model. Small output errors in the presence of sufficiently exciting inputs is often an indication of validity for a model so that it is not necessary to go further with the analysis.

The official statistical theory of identification proposes other parametric methods called OE, ARX, ARMAX, B&J, PEM, and so on [LJU 99]. Amongst these, OE (Output Error method) uses the least-squares method of output error above, in a statistical context. In the present case, it so happens that the conditions under which OE would be statistically optimal are not met: residual output should be a white noise, but this is far from being the case (Figure 7.6). In this event, the quest for optimality pushes us to make the model more complex and to jointly identify statistical parameters related to noises. Unfortunately, the results – which are not always convincing – depend on the type of model adopted (ARX, ARMAX, etc.) and there is no fully satisfying validation procedure. As a last resort, visual examination of the output error remains the decisive criteria. After testing various options, we decided to keep the OE; it has the benefit of staying away from any arbitrary choice, except for the selection of the structure of the model itself, without adding any further noise modeling.

The practical implementation of identification by output error is straightforward, providing the tools necessary for simulation and optimization are available. Therefore we will not elaborate further on the subject. There are also complete tools for identification, in particular the system identification toolbox from Matlab®. They generally provide an indication of the accuracy of the identified parameters, in the form of a variance matrix of estimation errors, which can only be correct if the optimality tests are positive, especially residual whiteness. If this is not the case, the structure of the model needs to be reviewed until satisfactory results are achieved. Although it is not optimal in terms of noise and effective disturbances, we will nevertheless remain with OE because of its robustness. Fortunately, this does not prevent the calculation of variance of estimation errors. Indeed, this does not pose any particular problems, but we describe it below due to its significance.

6.3. Estimating the error variance

We specify the dependence of the observed output y of a process in relation to vector θ of parameters, by writing $y = f(\theta) + v$, under the following matrix form:

$$\begin{bmatrix} y_1 \\ y_2 \\ \vdots \\ y_t \\ \vdots \\ y_N \end{bmatrix} = \begin{bmatrix} f_1(\theta) \\ f_2(\theta) \\ \vdots \\ f_t(\theta) \\ \vdots \\ f_N(\theta) \end{bmatrix} + \begin{bmatrix} v_1 \\ v_2 \\ \vdots \\ v_t \\ \vdots \\ v_N \end{bmatrix}, \text{ where } \theta = \begin{bmatrix} \theta_1 \\ \theta_2 \\ \vdots \\ \theta_n \end{bmatrix} \text{ is the vector of the parameters and}$$

where:

– $\{y_t\}$ is the series of observed outputs, from the initial instant $t = 1$ to the final one N;

– $\{f_t(\theta)\}$ is the series of the outputs which would result from the simulation of the exact model driven by the observed inputs;

– $\{v_t\}$ represents the noise and disturbance on the output, including those coming from input errors. This signal is assumed to be time invariant, ergodic and centered.

The OE estimate $\hat{\theta}$ is that which minimizes the criteria $J(\theta)=\|y-f(\theta)\|^2$. To carry out this minimization, optimization algorithms work through iteration, performing as many simulations as necessary. Around the optimum level, the first order development of the function $f(\theta)$ is written as:

$$f(\theta)\sim f(\hat{\theta})+F_\theta(\theta-\hat{\theta})$$

where F_θ is the Jacobian matrix of partial derivatives:

$$[F_\theta(t,j)]=\begin{bmatrix}\frac{\partial f_1}{\partial\theta_1} & \cdots & \frac{\partial f_1}{\partial\theta_n}\\ \vdots & \frac{\partial f_t}{\partial\theta_j} & \vdots\\ \frac{\partial f_N}{\partial\theta_1} & \cdots & \frac{\partial f_N}{\partial\theta_n}\end{bmatrix}$$

At optimum, the gradient of $J(\theta)$ in relation to θ is zero. It develops as:

$$\begin{bmatrix}\frac{\partial J}{\partial\theta_1}\\ \vdots\\ \frac{\partial J}{\partial\theta_n}\end{bmatrix}=2F_\theta^T(y-f(\hat{\theta}))=2F_\theta^T(v+f(\theta)-f(\hat{\theta}))=0$$

Replacing $f(\theta)$ by its development gives:

$$2F_\theta^T(v+F_\theta(\theta-\hat{\theta}))\sim 0$$

Define F_θ^+ as the pseudo-inverse of F_θ : $F_\theta^+=(F_\theta^TF_\theta)^{-1}F_\theta^T$. Pre-multiplying the equation above by $(F_\theta^TF_\theta)^{-1}$, gives the error estimation:

$$\hat{\theta}-\theta\sim F_\theta^+v,$$

The estimation of the error variance $V_{\theta\theta} = E[(\hat{\theta} - \theta)(\hat{\theta} - \theta)^T]$ is therefore approximately given by:

$$V_{\theta\theta} \sim E(F_\theta^+ v v^T F_\theta^{+T}) \sim F_\theta^+ V_{vv} F_\theta^{+T}, \text{ where } V_{vv} = E(vv^T)$$

To be able to use this expression, we determine the Jacobian matrix F_θ by finite difference, by carrying out *n* simulations successively varying each component: $\hat{\theta}_i \rightarrow \hat{\theta}_i + \delta\theta_i$. *According to the ergodicity hypothesis*, the variance matrix V_{vv} can be calculated using the autocorrelation function of the residuals $\hat{v} = y - f(\hat{\theta})$. This gives:

$$V_{vv}(i,j) \sim \frac{1}{N}\sum_{t=1}^{N} \hat{v}_t \hat{v}_{t+i-j}, \text{ where } \hat{v}_{t+i-j} = 0 \text{ when } (t+i-j) \notin [1,\ N]$$

6.4. Hypothesis test and confidence regions

Given the parametric vector θ of dimension *n*, a supposedly normal estimator $\hat{\theta}$ and its variance matrix $V_{\theta\theta}$, the variable *D* (*deviance*) $D = (\theta - \hat{\theta})^T V_{\theta\theta}^{-1}(\theta - \hat{\theta})$ complies with a law of χ^2 to *n* degrees of freedom. The confidence region around $\hat{\theta}$ with probability *P* is where *D* is below a threshold *S*, given by the tables of χ^2. As a result, for $P = 90\%$ and $n = 6$, tables give $S = 10.65$.

This means that we are able to carry out hypothesis tests. Therefore, to test the assumption $S_2 < 1.62$, we will perform the identification under this constraint, and calculate *D*, where $\hat{\theta}$ and $V_{\theta\theta}$ result from free identification, and θ from forced identification by the constraint. The hypothesis will be rejected at likelihood *P* if $D > S$.

We can also compare the relative likelihood of two hypotheses, by comparing the deviance *D* of two estimations in relation to the same reference created by free estimation.

If we extract from $\hat{\theta}$ and $V_{\theta\theta}$ a given parameter $\hat{\theta}_i$ and its variance $V_{\theta\theta i}$, the confidence levels are as follows:

$$66\%: \hat{\theta}_i \pm 0.954\,\sigma_i, \quad 90\%: \hat{\theta}_i \pm 1.645\,\sigma_i, \quad 95\%: \hat{\theta}_i \pm 1.960\,\sigma_i$$

For a pair of parameters, the inequality $(\theta - \hat{\theta})^T V_{\theta\theta}^{-1}(\theta - \hat{\theta}) < S$ defines the ellipsoid of confidence at:

$$66\%: S = 2.16 \quad 90\%: S = 4.60 \quad 95\%: S = 5.99$$

A hypothesis is rejected or not depending on whether the corresponding parameters are located outside or within the confidence region. We shall carry this out for sensitivities S_1 and S_2.

The degree of adequacy for the complexity of the model can be tested by observing whether a reduction in order, or the annulment of a given parameter, can be accepted or rejected. The validity of this detection method depends on the richness of the measurements available.

6.5. Conditions of application

The above calculation of variance (section 6.3) as well as its consequences (section 6.4), presume that the noises and disturbances which affects the input and output data have a minimum amount of statistical regularity, and that the duration of measurements is sufficient so that convergence towards normal laws can arise. When we are still far from convergence, calculations become more complicated, and the size of the confidence regions can increase in proportions which are difficult to evaluate.

Yet above all, this statistical regularity not only covers the hypothesis of *stationarity* (invariance in time, relatively well-verified by climatic archives), but also the hypothesis of *ergodicity*. The latter assumes that the alinement and calibration of the paleoclimatic reconstructions to modern data is correct. Indeed, in the event of an error in calibration, the sensitivity of the output in relation to the input would be irreparably effected by the initial calibration error, without the possibility of compensating this with an average over time, whatever the duration of the observation may be. The estimates and confidence regions are given with this reservation. If the conditions are not verified (and they are never perfectly so), it may be the case that, according to data used for identification, a single parameter may be assigned a number of quasi-certain region of confidence (99%) which are completely disjointed.

Furthermore, since we only have a limited number of reconstructions (four for temperature and four for irradiance), the number is insufficient for the law of large numbers to compensate this type of error, on which we have no *a priori* statistical characteristic (variance, independence). There would therefore be no sense in making averages from the estimates made, or in attempting to extract *likely* values or ranges of uncertainty. Indeed, alleged probability assessments would be worth no more than the scale of experts' judgments, which are bereft of any statistical basis.

In such a situation, one solution would be to present all the results of identification, each one being accompanied by a region of confidence and a comment (implicit or explicit): "*on the proviso* that the original data is correct"; i.e. without any noticeable errors of calibration or alignment. Everyone then remains free to use the results as they see fit and to draw their own conclusions from them.

7

Partial Results

7.1. A selection of data

One should recall that our collection of paleoclimatic data is made up of:

– four temperature reconstructions (Figure 2.3);

– one reconstruction of the CO_2 in the atmosphere (Figure 2.5);

– four reconstructions of solar irradiance (Figure 2.12);

– one reconstruction of volcanic activity (Figure 2.13).

Hence, there are 16 possible combinations, according to the temperature and irradiance chosen.

In this chapter, we have arbitrarily chosen the only combination from *Ljungqvist-HadCRUT4* with *Usoskin-lean0*, in order to present and give a detailed commentary on the results of identification and the simulation graphs. That does not mean we consider the databases selected as being the most reliable, nor that the model identified would be the most likely. We adopted this combination because it is the one which most clearly illustrates a number of points appearing in forms which are either too exaggerated or imperceptible in other combinations. In any case, we will not make any general conclusions from these first observations.

We must not only select the data, but also a period $[T_0, T_1]$ for the identification. The results of identification are sensitive to this and it is known that a truncation of data sometimes covers a manipulation intended to highlight hypothetical behavior, or to conceal it. Hence the explanations which follow.

Start of the period

All databases used do not go back to the year 1. Moreover, the oldest archives lack reliability: they present a dispersion which would lead to incoherent results if they were taken into account. Conversely, since the climatic past is sparse in terms of significant events, valuable information should not be sacrificed. In particular, it is important to fully cover the Medieval Warm Period, meaning a start date T_0 a reasonable time before the year 1000. We noted that overall the results fit together, all of them T_0 being situated between the years 500 and 1000. Two of the archives on irradiance happen to start in 695, which is why we have chosen this date for T_0.

End of the period

T_1 will be the current year in order to incorporate the latest available data. We will also carry out identifications stopped at $T_1 = 1999$, followed by blind short-term predictions in order to assess the predictive capacity of the models.

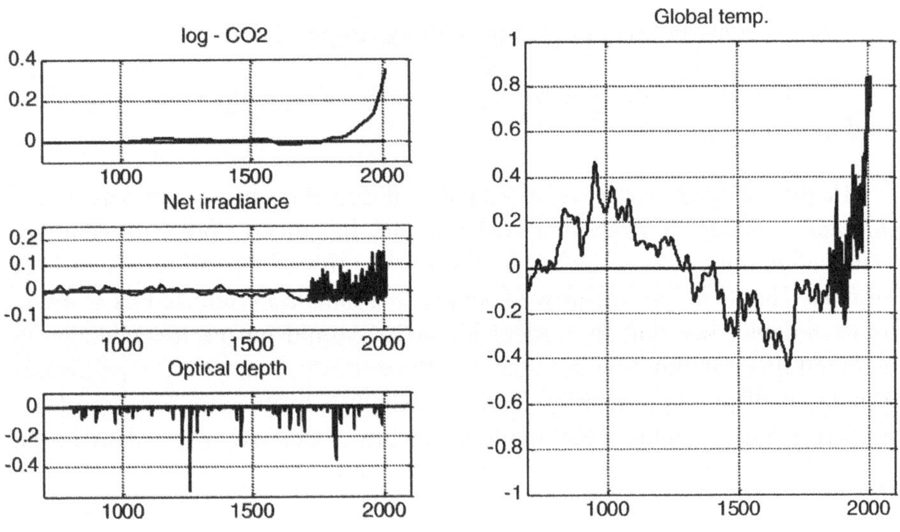

Figure 7.1. *An initial input/output set*

Figure 7.1 displays this data over the selected period: 695–2013. This is the raw, unsmoothed data as it is to be processed by the identification algorithms.

The noise which affects the modern measurements (after 1850) differ from earlier reconstructed temperatures. Nevertheless, they are accepted as they are without any arbitrary pre-processing. Likewise, irradiance reconstructions before 1700 filter the solar cycles. However, we can take advantage of the ones which are there. One of the benefits of OE is that it is robust in this type of heterogeneity.

7.2. Free identification

By *free* identification, it is understood that the parameters to be estimated are free of all constraints. The identification method (OE) also has no degree of manipulable freedom. The data therefore speaks freely, without being forced by any arbitrary assumption or *a priori* knowledge[1].

Simulations using the identified model

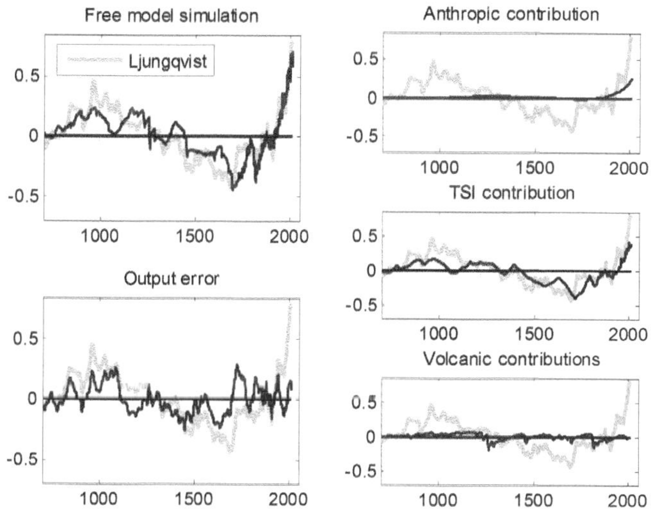

Figure 7.2. *Simulations from the free model*

1 Only the range of time constants [1, 10, 100 *year*] is somewhat arbitrary.

Immediately, we continue with the examination of the output simulated by the identified model, and the influence of various inputs on this output. In each graph in Figure 7.2, the grey line indicates the historic temperature observed. Every line is lightly smoothed for better legibility.

It can be noted that the influence of solar activity on recent warming is greater than that of CO_2. Volcanic activity plays a role which is marginal, but not negligible.

Output error is high. Deviations from one extreme to the other reach 0.5°C. These come firstly from natural random fluctuations of temperature, which cannot be explained by input variables (CO_2, *TSI* and volcanism). They also arise from uncertainties on measurements of paleoclimatic data, in both input and output. With identification, it is impossible to specify which explanation has the greater effect, but the examination of the period 1850 to now, where we have more reliable data, indicates that the influence of internal climatic variations is essential.

Even by making the model more complex and giving it more degrees of freedom, the error would not significantly change: the inputs do not have components which are correlated enough with the output over the whole duration of the identification. However, if we greatly restricted the period of identification, for instance to 1975 to 2000 (Figure 3.6), the output error over this period would be reduced to annual fluctuations, rather than reaching the ±0.2°C shown by Figure 7.2 over this period. In order to achieve this result, the identification algorithm would apply its six degrees of freedom by taking full advantage of the correlation observed during this period, between the rise in CO_2 and global temperature. The entirety of the warming observed during the second half of the 20th Century could then be attributed to the single anthropogenic factor. In reality, a period of 25 years is much too short to be sure that the correlation observed is not caused by random climatic fluctuations: this would be a blatant example of "cherry picking" performed solely to support the argument of a given hypothesis. To a lesser extent, this could also be applied to simulations limited to the period 1850 to now, which avoids going back to the LIA and the MWP.

Parameters identified

Minimizing the output error leads to the following parameter estimations:

S_{clim} (°C) Climate Sensitivity	S_{irrnet} (°C/Wm^{-2}) Sensitivity to irradiance	S_{vol} (°C) Volcanic activity	T_{clim} (an) Climate time constant	S_{irr11} (°C/Wm^{-2}) Frequential sensitivity	T_E (°C) Equilibrium temperature
1.28	**17.5**	10.7	90.0	0.90	0.027

NOTE.– It should be recalled that $S_{clim} = S_1 \log(2)$. The sensitivity to net irradiance can be noted equally as S_2, S_{irr} or S_{irrnet}, and S_{irr11} is the sensitivity at the frequency $f = 1/11$ (see 4.4).

The climate sensitivity found (1.28°C to a doubling of CO_2) is compatible with the IPCC's assumed window (1 to 6°C). However, the sensitivity to irradiance calculated (17.7°C/Wm^{-2}) is ten times higher than the upper limit, ($S_2 < 1.6216°C / Wm^{-2}$), which was reported as *very likely* (i.e. at 90%).

The values identified are in relation to the respective influences of anthropogenic and natural factors in Figure 7.2.

Static and dynamic sensitivities

The climatic time constant identified is 90 years. Given the thermal inertia of the oceans, that is understandable. This means that the climate continues to react to past influences during a significant length of time, which does not facilitate the visual interpretation of climatic causality. Yet the model reacts almost instantly to pulse inputs, as is shown by the response to each volcanic eruption. This behavior characteristic of thermal processes demands a model structure with several time scales like the one we have adopted, without limiting ourselves to a simple first order. In fact, the step response of the model identified provides a similar appearance to that of Figure 4.2, representative of the GCM simulation. The difference between this and a simple first order (dotted line) is minor, but it plays a key role in the immediate reactions of the climate to solar or volcanic influences.

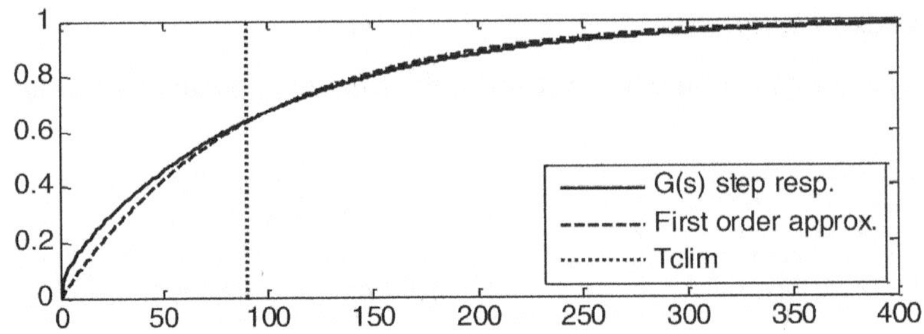

Figure 7.3. *Dynamic response*

Indeed, for the contribution of irradiance, over the period where it is present in the data, a response to the Schwabe cycles can be noted. Figure 7.4 (left) overlaps the temperature observed between 1960 and 2010, corrected for estimated human activity, volcanism and an underlying trend, with the estimated contribution of solar activity. The right graph shows similar lines, obtained by C. Camp *et al.*, [CAM 07], in the context of determining sensitivity to irradiance using modern measurements.

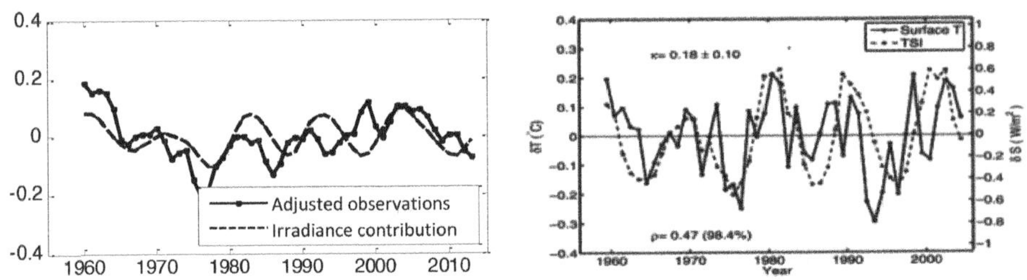

a) b)

Figure 7.4. *Reproduction of solar cycles by the contribution of irradiance*

The temperatures in the two graphs above overlap poorly because the respective processes of smoothing and removal of trends differ. Camp estimates the sensitivity at $0.18 \pm 0.10°\mathrm{C}/Wm^{-2}$, or $1.03 \pm 0.57\,°C/Wm^{-2}net$. Our identified frequential sensitivity is very similar: $S_{irr11} = 0.9\ °C\ /\ Wm^{-2}net$. This coincidence is noteworthy, given that our estimation is not the result of processing the historical period selected by

Camp for this analysis, but from data covering more than a millennium, that dynamic parameters are common to all causality chains, and that their identification results from a processing of the collection of all data, including CO_2 and volcanism.

However, there is a difference in the nature of these two "sensitivities". Ours corresponds to a given point of the frequential response of the climate process (frequency 1/11 *years*), and not only manifests itself by a gain, but also with a phase change, which here is of 47°. The correlation established by Camp regards values without phase changes, which would only make sense for a static model (reduced to $G(s)=1$), or for signals which vary very slowly. The result is that the response to the Schwabe cycles does not prejudge any level of static sensitivity. This is much higher than the dynamic sensitivity, given the ratio $\rho = 0.9/17.5 \sim 1/20$. Camp's estimation technique (*composite mean difference projection*) is not in question here: it is the object of his estimation itself which is inconsistent. Likewise, the temperatures in Figure 3.1 (the hockey stick as well as boomerang curves) do not vary sufficiently slowly in relation to the time constants which are of the order of a century. In fact, the only significant "correlations" of a cause and effect relationship are those which are expressed via the identification of a dynamic causal model. This is the entire purpose of this work.

Finally, apart from the fact that sensitivity to irradiance and its contribution to warming are incompatible with the IPCC's assumptions, there is nothing wrong to be added about the simulation as a whole. For instance, it can be noted that the estimated equilibrium temperature is T_E = 0.027°C, close to the reference of zero from 1850–1900, the assumed pre-industrial equilibrium temperature.

7.3. Forced identifications

Contrary to "free" identification, here the model is now constrained to comply with arbitrary assumptions on given parameters. The parameters which remain free are adjusted to optimize temperature reproduction.

Constraint on sensitivity to solar irradiance

We now identify, under the constraint of the sensitivity assumed by the IPCC, related to solar irradiance: $S_2 < 1.62°C / Wm^{-2}$. In this way, we can obtain the following set of parameters, and the simulations in Figure 7.5.

	S_{clim} (°C)	S_{irr} (°C/Wm^{-2})	S_{vol}(°C)	T_{clim} (an)	S_{irr11} (°C/Wm^{-2})	T_E (°C)
Free model	1.28	17.5	10.7	90.0	0.90	0.027
Forced model	3.92	**1.62**	21.1	97.9	0.05	-0.020

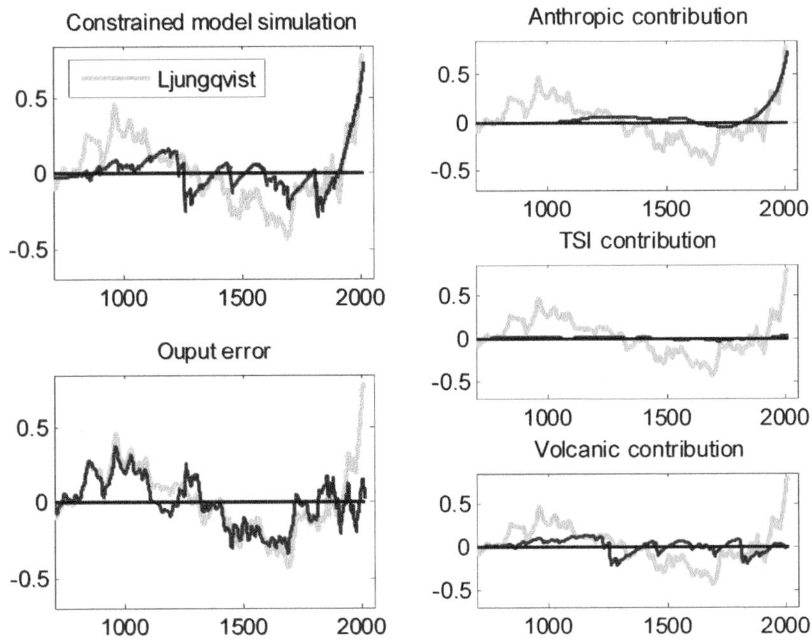

Figure 7.5. *Simulation forced model*

Due to the constraint on solar irradiance, the warming seen on the 20th Century is then attributed almost completely to CO_2. Despite a doubling of the coefficient on the sensitivity to volcanic activity, and a reduced equilibrium temperature, the temperature simulated over the entire millennium is far from matching the one that was produced by free identification. Figure 7.6 compares the output errors with and without the constraint on sensitivity to irradiance, and shows ranges at ± 1 standard deviation.

The deterioration in output error (with 35% variance) is blatant. This is largely due to the poor reproduction of preindustrial temperatures, caused by a lack of sufficient contribution of solar activity. Through the optimization method, the estimated rise in climate sensitivity S_{clim} (from 1.28 to 3.92°C) nevertheless allows the error to return to zero more quickly after the year 2000.

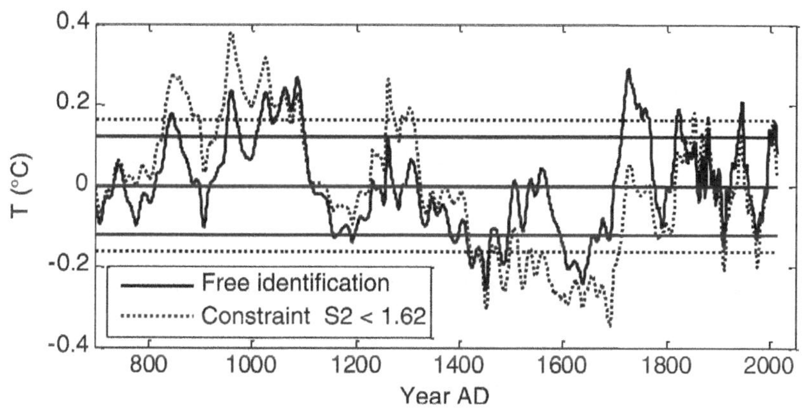

Figure 7.6. *Comparison of output errors*

Moreover, the weak static sensitivity coefficient imposed on solar irradiance does not allow us to obtain sufficient dynamic sensitivity, despite the parameters remaining free in the model. The contribution of Schwabe cycles on modern temperatures is significantly affected, as shown in comparative Figure 7.7.

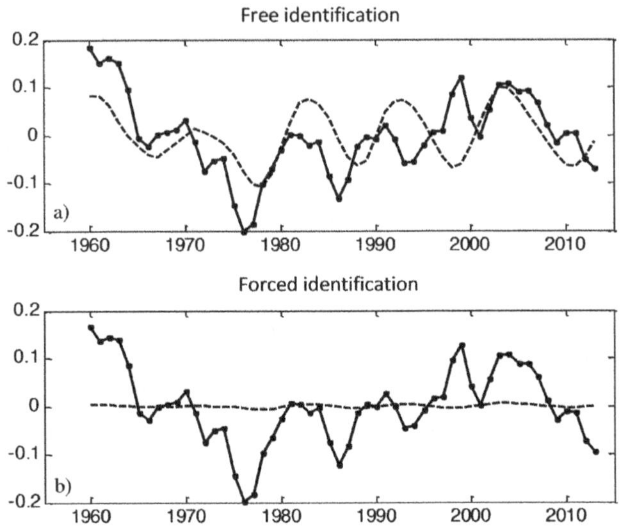

Figure 7.7. *Reproduction compared with Schwabe cycles*

In conclusion, the model with a constraint on solar activity is, at first glance, much less convincing than the freely identified model. This must now be confirmed by statistical analysis, and especially by results from other data combinations.

Constraint of nullity for anthropogenic sensitivity

Finally, identification under the (unlikely?) hypothesis of insignificant anthropogenic sensitivity can be carried out. With the constraint $S_{c\lim} = 0$, we obtain the following collection of parameters, and the simulations shown in Figure 7.8.

	S_{clim} (°C)	S_{irr} (°C/Wm^{-2})	S_{vol} (°C)	T_{clim} (*an*)	S_{irr11} (°C/Wm^{-2})	T_E (°C)
Free model	1.28	17.5	10.7	90.0	0.90	0.027
Forced model	**0**	19.5	8.7	86.5	1.09	0.04

With the obvious exception of S_{clim}, estimates remain close to those seen with free identification.

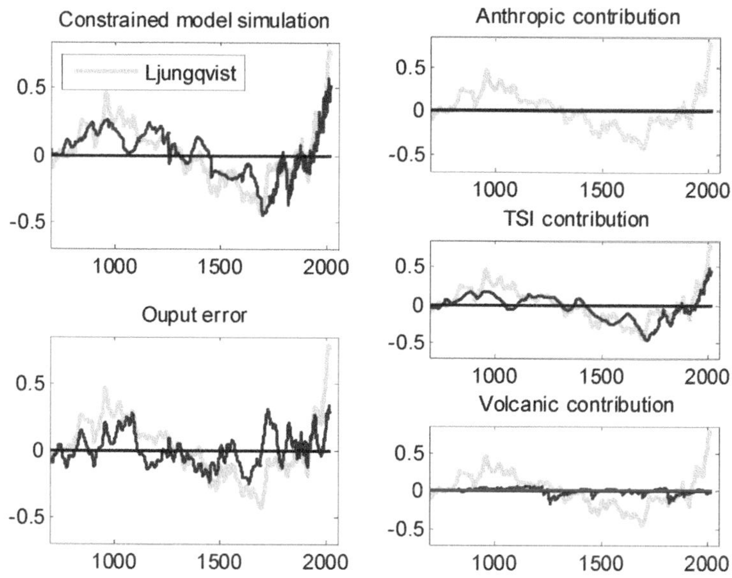

Figure 7.8. *Hypothesis of zero anthropogenic contribution*

Figure 7.9 shows the output errors with and without the constraint $S_{clim} = 0$, and the ranges at ± 1 standard deviation.

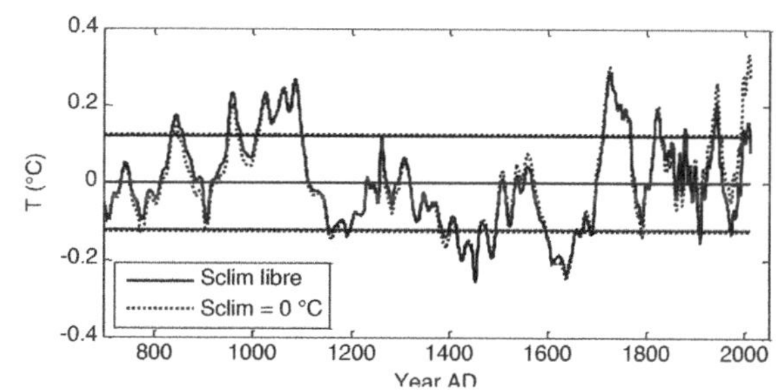

Figure 7.9. *Comparison of output errors (S_{clim} = 0)*

The increase in the output standard deviation is hardly perceptible (2%), except around the year 2000. In effect, we are deprived of the degree of freedom regarding the CO_2 input, which is active at precisely this point. Nevertheless, the corresponding error is not excessive compared to the output deviations over the past millennium. In other words, the hypothesis that the warming between 1970 and 2000 is entirely due to solar activity and natural climatic variation does not seem incompatible with the observations. Once again, this needs to be confirmed by statistical analysis.

Constraint on sensitivity to volcanic activity

A detailed illustration of this point would not be of any great value. If the sensitivity to volcanic activity is constrained to zero, the degradation of the output error is insignificant.

7.4. Statistical analysis

Output errors

Let us now reconsider the output errors in Figure 7.6 (with and without the constraint on sensitivity to TSI). Despite the temporal variability of paleoclimatic data (proxies, the instrumental measurements), they do provide a certain regularity, which justifies the examination of correlation and inter-correlation functions. The difference between the two types of errors can be clearly seen on the graphs in Figure 7.10.

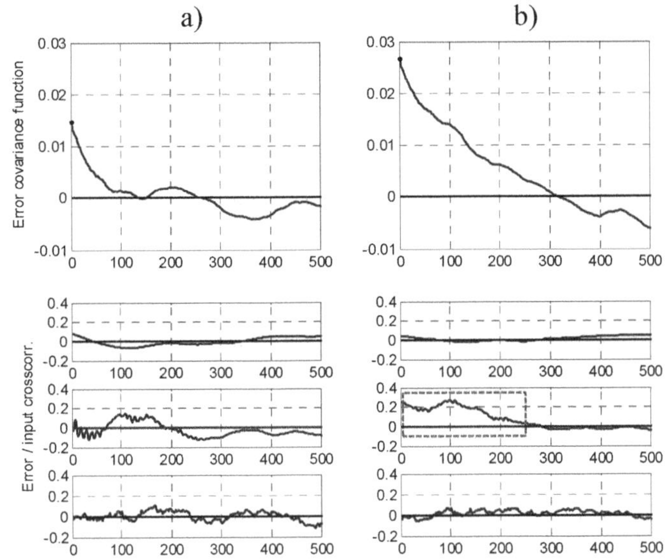

Figure 7.10. *Functions of auto and cross-correlation of the error: free identification a) and identification forced by $S_2<1.62$ b)*

At the top, the functions of autocovariance have been traced. The initial dot shows the value of the variances. It is necessarily greater in the case of forced identification (right), where the minimization lacks the degree of freedom S_2. The question is to know whether this increase is significant. In both cases, the error is far from being a white noise (for which the autocorrelation function would be an impulse), but the "coloring" implied by the wider range is more pronounced for the forced identification. And above all, the presence of a residual cross-correlation with the irradiance input (in a dotted-line box) shows the causality which the constrained model does not sufficiently account for.

Estimated confidence intervals

For all of the following, the confidence intervals and regions associated to the identified parameters are given at 90% (*very likely*, according to the IPCC terminology). This corresponds to confidence intervals at ± 1.645 standard deviation.

For climate sensitivity, we have, in accordance with the calculation methods in section 6.2:

$$S_{c\lim} = 1.28\ [-0.53 \text{ à } 3.09]°C \quad (90\%)$$

The confidence interval covers the value $S_{c\lim} = 0$. It confirms that the hypothesis of zero climate sensitivity is not incompatible with the combination of paleoclimatic data used.

The upper value of the confidence range obtained is much lower than the extreme limit (6°C) stated by the IPCC. This value is unrealistic given the data processed.

These results are already significant, but the range is too wide for the estimated value to constitute a reliable indicator.

For sensitivity to net irradiance S_2 (also noted as S_{irr}), we have:

$$S_{irr} = 17.5\ [10.6 \text{ à } 24.4]°C / Wm^{-2} \quad (90\%)$$

This time, we have a robust indication: the low sensitivity to solar activity stated by the IPCC ($S_{irr} < 1.62°C / Wm^{-2}$) is strictly incompatible with climatic observations. The subjective impressions in section 7.3 on the inadequacy of the constrained model were therefore not an illusion.

Confidence regions

Here we consider the joint confidence regions for the pair of parameters (S_{clim}, S_{irr}). Regions at 90% are defined by the inequality $(\theta-\hat{\theta})^T V_{\theta\theta}^{-1}(\theta-\hat{\theta}) < S$, where S=4.60 corresponds to the probability threshold at 90% of a variable of χ^2 with two degrees of freedom. Figure 7.11 shows both the confidence region of 90%, from identification (unbroken line), and the *a priori* assumed level (grey rectangle). Strictly speaking, the latter would need to be refined by an ellipsoid taking into account the interdependence between the two sensitivities S_{clim} and S_{irr}, since both are connected to the uncertain parameter S_G. In any case, even taking into account the ambiguity which characterizes the IPCC's probabilities, the identified and assumed level are too far from each other for there to be any possible intersection.

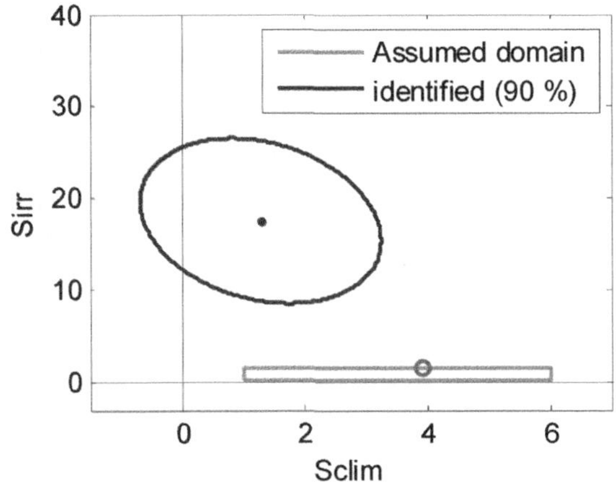

Figure 7.11. *Assumed and identified regions*

The couple (S_{clim}, S_{irr}) identified under the constraint $S_2 < 1.62°C\ /\ Wm^{-2}$ is shown by the mark **o.** The deviance associated to this model is:

$$D = 38.3$$

This is much higher than the threshold $S = 10.64$. The model constrained by the IPCC assumption must therefore be rejected with a probability of 90%.

It should be recalled that the results of this chapter only relate to a single combination of paleoclimatic data. The conclusion above is therefore subject to the confidence given to the data in question.

8

Overall Results

8.1. Preliminary comments

We will now combine the four temperature reconstructions (in order: Moberg, Ljungqvist, Loehle, Jones-Mann, all spliced with HadCRUT4), with the four irradiance reconstructions (in order: Usoskin-lean0, Usoskin-timv15, Be10-lean0, Be10-timv15).

Some results in this chapter will pose interpretation problems which should be anticipated. Indeed, it may be of some surprise that some confidence regions and intervals show an empty intersection, both between one another and with the IPCC assumptions. As such, it will be noted that the identified sensitivities to solar irradiance are mutually incompatible, depending on whether the reconstructions are based on lean0 or tim15. More specifically, the regions of confidence from Jones & Mann will be systematically remote or disconnected from all the others. Even though these are confidence regions of only 90%, and not 100% certainties, we are led to the conclusion that certain data, hypotheses or conclusions are incompatible, without being able to ascertain which are correct and which are not.

Let us look back here at the notion of confidence regions. In identification, they are traditionally determined using the input/output data itself, and not deduced from *a priori* statistical characteristics, like variance in measurement errors. Indeed, these characteristics themselves are identified, in this case, through the function of autocovariance of the output error. This is why we have systematically ignored uncertainty indicators which often accompany the collected data.

Moreover, it is traditional practice to use different series of measurements, occasionally of various qualities. As a general rule, they are implicitly assumed to be valid: not affected by bias or significant calibration errors. If such defects are present, uncertainty calculations on identified parameters no longer hold, due to a lack of ergodicity (condition mentioned at the end of paragraph 6.3). Indeed, the major difficulty encountered in the paleoclimatic reconstructions regards the calibration of the proxies used. When the calibration is incorrect, due to the choice of proxies or the method of statistical processing, an error is introduced which cannot be compensated for by average over time, carried out through the identifications.

Prior to the technical hypothesis of ergodicity, it goes without saying that the identification data is supposed to reflect the same process. In this case, this would occasionally seem rather doubtful. The hockey stick curve, which ignores large climatic events, seems to have come straight from another world; one of a pre-industrial lost paradise often mentioned by Sir John Houghton (Figure 3.1). Similarly, the ER background corrections are explicitly inspired by extrasolar systems.

Finally, if the reconstructions lead to disconnected regions of uncertainty, we can only take note of this and reject all or part of the conclusions derived, either from one or from the others. This selective rejection must therefore be based on criteria other than those of identification, for instance the level of confidence (*low, medium, high*) given to one type of reconstruction with regard to another. But who is empowered to make such distinctions?

Moreover, the uncertainties calculated are only approximations, asymptotically valid when the identification durations tends to infinity, which is relatively far from being achieved. They also assume the time invariance of statistical characteristics of noises and measurement errors, which are significantly different, depending on whether modern measurements or paloeclimatic measurements are being used. In the end, even if the thresholds and confidence intervals only have approximate values, we accept that they constitute more objective assessment criteria than the assertion of confidence from experts, assured only by their own personal judgment.

The analysis which follows therefore gives the elements for the systematic evaluation of the results, leaving the reader free to decide upon their own levels of confidence as they see fit.

8.2. Regions and intervals of confidence

Figure 8.1 shows the identified confidence regions associated with the 16 data combinations, grouped according to paleotemperatures. It is almost redundant alongside the following Figure 8.2, where, to within one factor, the intervals of confidence are given by the width and height of each ellipsoid.

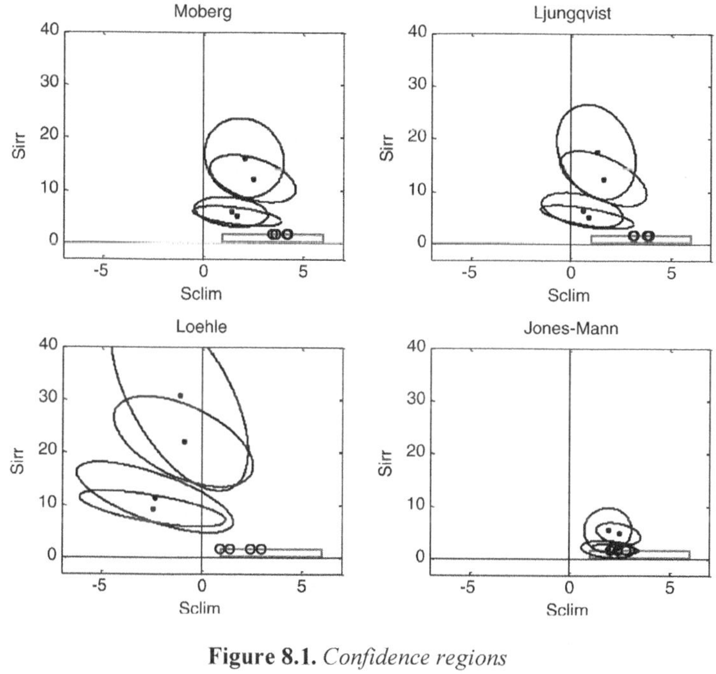

Figure 8.1. *Confidence regions*

Given the diversity and the uncertainties which affect the paleoclimatic data, we should not be surprised that the estimates are significantly scattered. Some would probably like to conclude from this that the principle of modeling by identification is fruitless, and unable to lead to quantified estimates of fundamental climatic parameters. Nevertheless, using the results above, we will extract a substantial quantity of information.

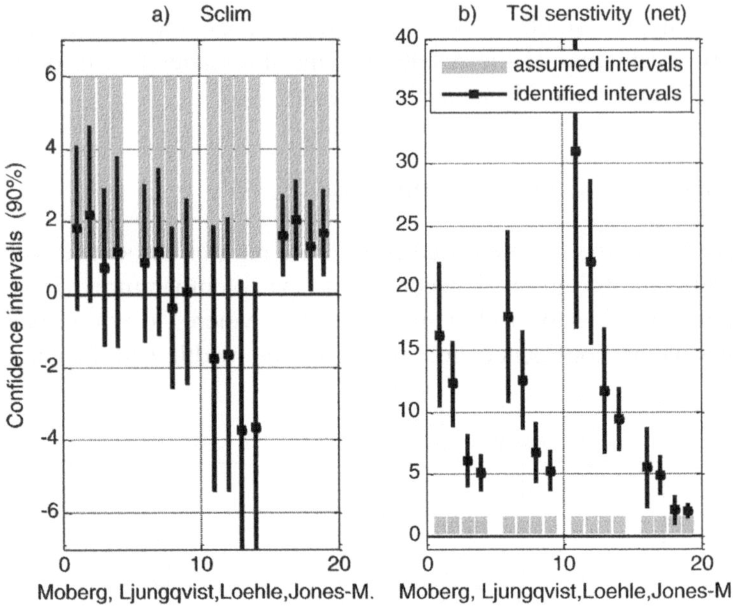

Figure 8.2. *Intervals of confidence (90%)*

It is shown by Figures 8.1 and 8.2 that the confidence regions relating to Jones and Mann are the most reduced. This can be explained by the fact that the amplitude of the millennium-old variations of the *Jones and Mann* reconstruction is the weakest of the four. Hence, mechanically, the output errors are the lowest, and ultimately, the estimated variances are the smallest. The opposite reasoning also explains the wider uncertainties from the Loehle reconstruction. Pushing this reasoning to the limit, a hockey stick curve with a perfectly rectilinear handle would lead to ranges with a width of almost zero, which this would clearly not validate the methods used to obtain this curve.

Figure 8.2 shows that the ranges of confidence estimated for climate sensitivity are generally incompatible with those of the IPCC, apart from those which result from the Loehle reconstruction. It also appears that the upper limit of the IPCC's assumed range ($S_{c\lim} = -6°C$) is never reached by the ranges identified, even when they come from Jones and Mann. According to their data, the lower and upper limits for climate sensitivity should be reduced to $[1-3°C]$.

Above all, it can be noted that the estimated sensitivities to solar irradiance are up to ten or twenty times higher than the maximum values assumed by the IPCC (1.62 $°C/Wm^{-2}$). For instance, let us examine the range of estimates which result from Ljungqvist (Figure 8.2(b)). The 5th and 6th segments result from the two TSI proxies spliced with *lean0*, and the 7th and 8th with *timv15*. The two pairs estimated are in a ratio of around 1:3, and the ranges are mutually incompatible. The existence and the level of the ER background correction are therefore in question, and we have no recommendation to make in this regard.

Finally, two points can be drawn out.

– Firstly, of the 16 combinations, only the final two, both based on Jones and Mann, and only under the condition of ER background correction, indicate a sensitivity to solar activity which is compatible with a sensitivity reduced to the energy factor only;

– Secondly, the selection of the irradiance reconstruction only has a small effect on the estimate of climate sensitivity S_{clim}, as well as on the real contribution of solar activity to recent warming. Indeed, Figure 8.3 shows the contributions related to the two Ljungqvist combinations with *lean0* and *timv15*. Despite the sensitivities of S_{irr} being in a ratio of 3:1, the orders of magnitude of the contributions are the same, thanks to amplitudes of irradiance reconstructions being in the inverse ratio.

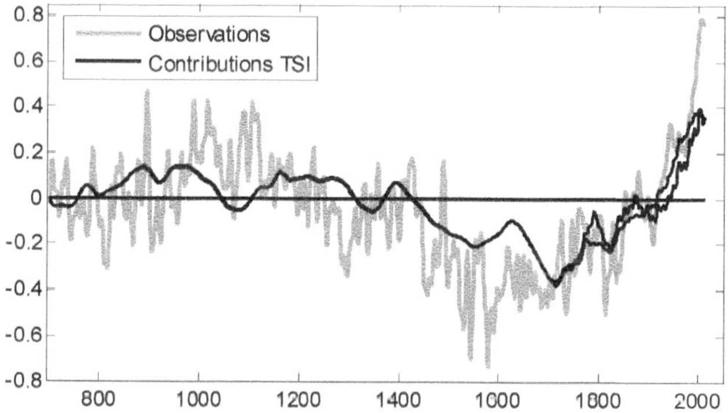

Figure 8.3. *Sensitivity of contributions of solar activity to irradiance reconstructions*

At this stage, the existence of a mechanism of solar activity with a powerful influence over the climate seems irrefutable, unless we reject any paleotemperature which is significantly different from the Jones and Mann reconstruction.

8.3. Hypothesis test

Hypothesis of solar sensitivity restricted to the energy factor

We have seen (see section 5.5) that this hypothesis is expressed by the constraint $S_2 < 1.62$. Figure 8.1 shows (marked **o**), the couples S_1, S_2 estimated under this constraint. The table of deviations between the constrained models and the freely identified models is the following (rows: temperature reconstructions, columns: irradiances):

$$D = \begin{bmatrix} \mathbf{21.7} & \mathbf{26.7} & \mathbf{13.5} & \mathbf{16.3} \\ \mathbf{38.3} & \mathbf{20.6} & \mathbf{16.1} & \mathbf{14.8} \\ \mathbf{439.8} & \mathbf{32.5} & \mathbf{344.7} & \mathbf{35.1} \\ 5.7 & 12.9 & 0.2 & 1.1 \end{bmatrix}$$

The rejection threshold for a model at 6 degrees of freedom is $D = 10.6$. This table therefore confirms the conclusions of the previous paragraph: the hypothesis of a low sensitivity to solar activity must be rejected as being incompatible with paleoclimatic reconstructions, apart from that of Jones and Mann (in the final line).

Hypothesis of zero climate sensitivity

According to the IPCC, a climate sensitivity below 1°C is *extremely unlikely*, and even more so, a sensitivity of zero. Nevertheless, we have tested this hypothesis $S_{clim} = 0$. We should recall that the climate sensitivity identified is the sensitivity to the concentration of CO_2, considered as a global indicator of human activity. It combines the climate sensitivity in its strictest sense (to CO_2), with sensitivities to other GHGs, industrial aerosols and changes in land use. In the event where negative effects are greater than positive forcings, the anthropogenic sensitivity could become negative. The table of deviations associated to $S_{clim} = 0$ is the following:

$$D = \begin{bmatrix} 6.0 & 7.7 & 2.9 & 3.0 \\ 3.7 & 4.5 & 0.4 & 0.8 \\ 1.6 & 0.4 & 6.6 & 3.1 \\ \mathbf{21.4} & \mathbf{31.7} & \mathbf{14.6} & \mathbf{23.9} \end{bmatrix}$$

As could be expected, a zero anthropogenic sensitivity is compatible with the first three reconstructions, and incompatible with that of Jones and Mann.

8.4. Comments

The fundamental characteristic of our approach of identification is that once the input-output data is selected, no degree of freedom is left, meaning there is no possibility of the conclusions being oriented in one direction or the other. The constraints of forced identification are openly implemented, in order to make the analysis process completely explicit. In this context, it is impossible to reject a given reconstruction for the benefit of another. It is also impossible to carry out average estimates to obtain the "*most likely*" model. Moreover, it is impossible to propose ranges which would result from the combination of regions which sometimes lack a common intersection. We therefore concur with the IPCC's attitude whereby, in the absence of consensus, no "*best*" climate sensitivity is assigned, and we remain with the presentation of the raw results, leaving the reader the freedom to draw their own conclusions. We will maintain this position in the following chapter on climate projections. We can only showcase the overall results obtained, allowing the reader to interpret them according to the confidence they have in the various reconstructions and their own preconceived ideas, whether they are skeptical or alarmist.

An initial conclusion is that the Jones and Mann reconstruction is the only one which is coherent with the IPCC's assumed climatic parameters. If we accept this as valid, the other three would have to be rejected, and vice versa.

In the event where the latter three are kept, we can reach two types of conclusion. The first is positive: solar activity not only explains the previous large climatic episodes, but also makes a strong contribution to the warming over the last century. Whatever the case in point, the relative accuracy with which sensitivity S_{irr} is estimated leaves no space for any doubt.

The second conclusion is less clear-cut. It comes down to stating that the observations do not allow us to determine climate sensitivity with any great confidence, not even the sign. Indeed, some free models display estimates of negative climate sensitivity, even significantly so (up to –2.46°C). For such values to be reached, the positive effect of anthropogenic GHGs would have to be much lower than the IPCC assumptions, which cannot be ruled out, but also, the negative effects of aerosols and changes in land use would have to reach extreme level, which is more doubtful.

9

Historic Minuscule Simulations

9.1. Overview of IPCC simulations

The term "historic" refers to the period 1850 to the current year, where simulations can be compared to instrumental measurements, available from this period over almost all parts of the globe.

Figure 9.1 below is taken from the fifth *Assessment Report* (AR5, Figure 10.21). It gives a graphic summary of multiple results obtained from the *Coupled Model Intercomparison Project* (*CMIP*5). This summary's main author is Phil Jones [JON 13], Director of the CRU, already mentioned several times above.

Three lines can be seen in each graph: 1. Historic observations; 2. The average and the range of simulations only using natural forcing; 3. The average and the range of simulations using both natural and anthropogenic forcing. The uncertainty ranges are claimed to be at 90%. Despite the problems of legibility, it can be noted that the natural factors alone are far from able to account for the increase in temperature, or in Ocean Heat Content (OHC).

To achieve this result, dozens of research centers independently developed very large numerical models and carried out numerous, long and costly simulations. In light of this, a study as limited as ours would seem not to have the same clout. However, we will present alternative (and incompatible) simulations, and propose some explanations for them.

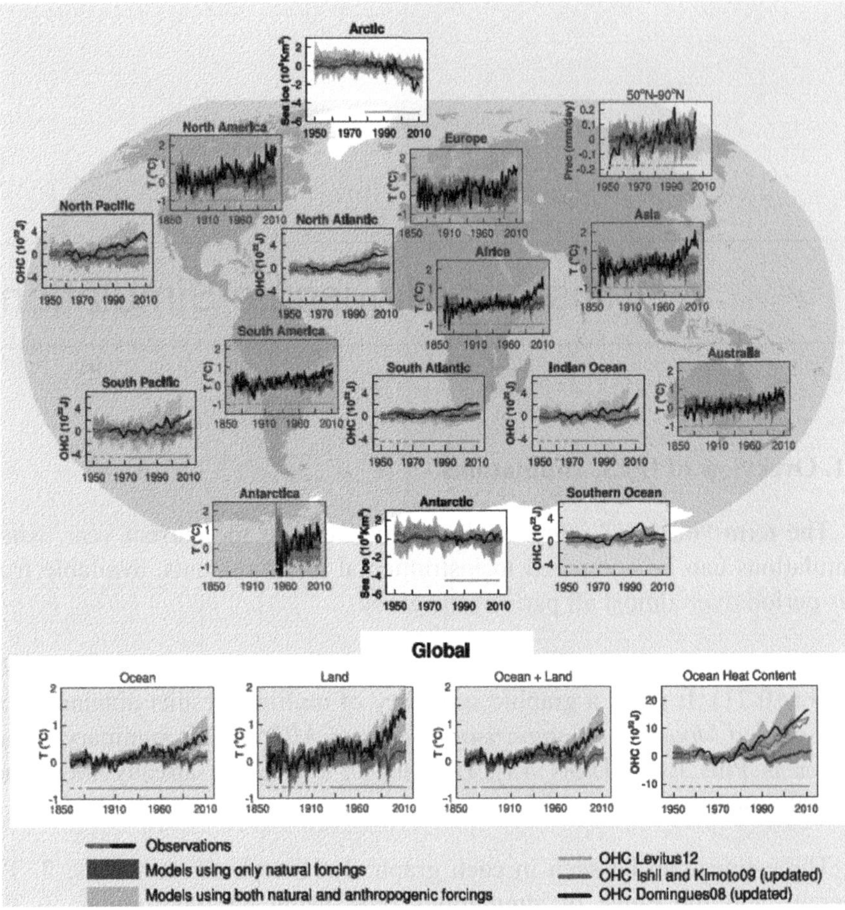

Figure 9.1. *IPCC simulations*

9.2. Comparative simulations

We are only interested in global temperature. Figure 9.2 therefore takes the "Ocean + Land" graph from Figure 9.1.

Figure 9.3 below shows the averages of the simulations, taken from all of our freely identified models, using all of the combinations in our catalogue of climatic data, without excluding the reconstructions of both Jones and Mann, and Loehle. This figure arises from the entirety of the data without any kind of external intervention.

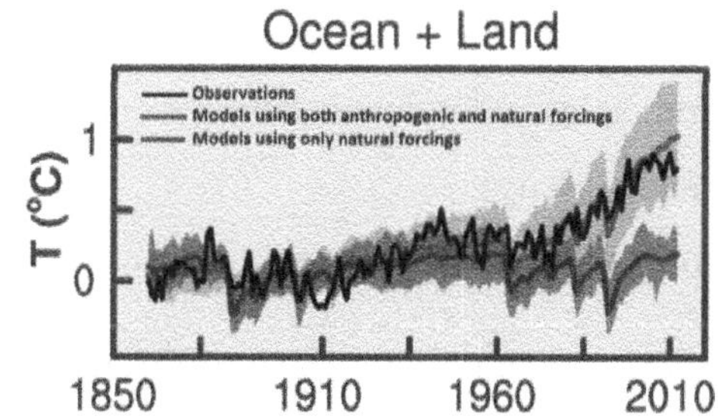

Figure 9.2. *CMIP5:Complete observations and simulations (upper lines), Natural forcing only (lower line)*

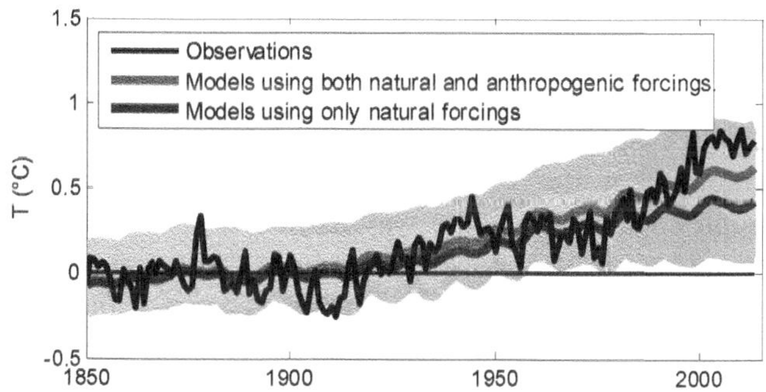

Figure 9.3. *Simulation of identified models (period 1850-2013)*

Both figures show the averages of the simulations, respectively with and without anthropogenic contribution. The conclusions are incompatible: for the IPCC, warming is almost entirely attributed to the anthropogenic factor, with the natural factors being dominated by volcanism. If the identified models are to be believed, the majority results from natural factors, and more specifically, from solar activity.

More precisely, for a warming of around 0.8 degrees since the start of the 20th Century, a visual evaluation gives the following distribution (in 2014).

According to the IPCC "knowledge-based" models:

+0.2°C results from solar and volcanic activity;

+0.8°C from human activity;

–0.2°C from natural random fluctuations.

According to observations (identified models):

+0.4°C results from solar and volcanic activity;

+0.2°C from human activity;

+0.2°C from natural random fluctuations.

These values are simply guideline ones and the spreads indicated in the graphs are of a different nature.

In our case, the dispersion results from that of the paleoclimatic reconstructions used, both for the identification and the simulations, and we will not extract ranges of probability or debatable confidence indices from them. It is important to specify that, even though the display period adopted is limited to 1850–2013, all of the millennium-old data contributes to the result.

In light of this, the variety of IPCC simulations results in part from the internal variability of the climate simulated that large coupled models are able to reproduce, in the same way as numerical wind tunnels simulate the turbulence of real flows. It also comes from the dispersion of coefficients of the climatic models themselves, in this case, those of the coefficients and parameters of the models which are not determined by the rigorous laws of fundamental physics (fluid mechanics, thermodynamics, etc.) The relatively small amount of dispersion in the results therefore would appear to reflect the consensus of numerous independent experts on the least certain characteristics of the climate machine. We will see that this is not the case.

9.3. *Representative concentration pathways (RCPs)*

The simulations to be carried out in light of the 5th IPCC report were defined in 2008 by the Working Group on Coupled Modelling (WGCM) of the World Climate Research Program, in the context of the CMIP5: Coupled Model Intercomparison Project, Phase 5, [TAY 12]. The scenarios regarding human activity are called RCPs: Representative Concentration Pathways [MEI 11].

The RCPs replaced the emission scenarios of previous projects. The essential new point is that the RCPs offer scenarios directly in terms of radiative forcing, and not in terms of emission or concentration of GHGs, aerosols, land use and other consequences of human activity.

However, the RCPs do not only address the future, they dictate to the participants of the CMIP5 the reconstruction of past radiative forcing over the historic period.

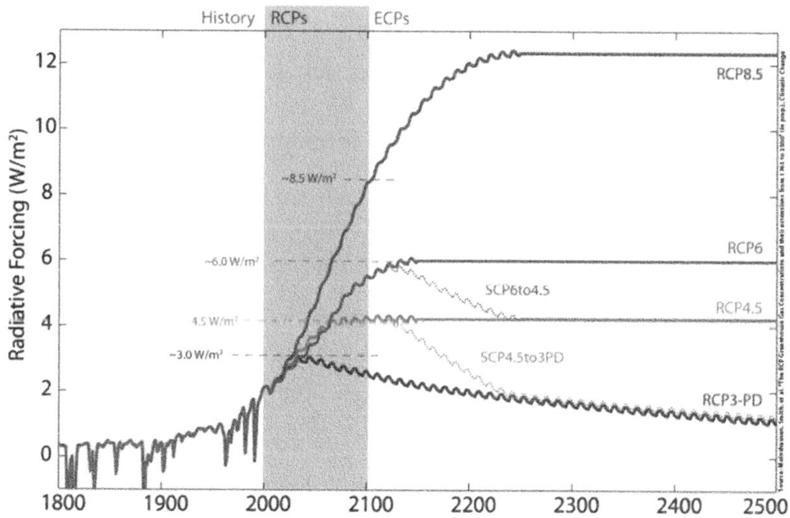

Figure 9.4. *The IPCC's RCPs*

Under the heading "History", the left part of the lines in Figure 9.4 show this reconstruction, which we will be examining here. We will return to the right section regarding long term climatic projections in the following chapter.

To focus on the nature of the RCPs itself: they describe a global radiative forcing, in Wm^{-2}. In other words, the RCPs integrate the model of all the machinery of radiative forcing: the greenhouse effect mechanisms, albedo variations, volcanic activity and solar activity reduced to the energy factor. The global value indicated is the sum of all the reconstituted or anticipated radiative forcing from day to day over the Earth's surface, to within a grid of 0.5 degrees latitude and longitude. The input data supplied to the CMIP5 participants is made up of the huge corresponding database, ready to be simulated by atmosphere-ocean coupled general circulation models.

These models comply both locally and globally with the fundamental principle of heat conservation: there can be no "missing heat" and, in the end, their global behavior complies with that of the elementary energy balance models. Supplying the flows of radiative forcing therefore equates to supplying an entire part of the model, corresponding to the sum of $Q = \alpha_1 u_1 + \alpha_2 u_2 + \alpha_3 u_3$ present in the linearized model $I_G(s)\, s\, x = \alpha_1 u_1 + \alpha_2 u_2 + \alpha_3 u_3 - \lambda_G x$ (see section 4.1).

Under examination, the "historic" forcing in Figure 9.1 appears to effectively consist of three components:

1) A basis of anthropogenic growth, approximately exponential over the historic period, corresponding to the term $\alpha_1 u_1$;

2) The solar contribution $\alpha_2 u_2$ results in small and barely visible oscillations for periods of 11 years, caused by solar cycles. The relative weakness of the coefficient α_2 makes the slow variations of TSI undiscernible;

3) The negatives peaks associated with volcanic eruptions.

In reality, the label "history" at the top of Figure 9.1 is no more justified than the alleged determination of radiative forcing by observations (see section 5.8: AR5 8.3). Only the observations in the strictest sense are historic: concentrations of GHGs, TSI, AOD, etc. The assumed radiative forcing is not.

So what degrees of freedom are left to the participants of the CMIP? The inertias and thermal exchanges are determined by the laws of physics: thermodynamics and fluid mechanics. Therefore, this essentially leaves the (distributed) coefficient of climate feedback λ_G, where the clouds play the central role, through their properties of radiation, emission, absorption and reflection. Above all, prior to this, the mechanisms governing the genesis of cloud formations. According to the IPCC, "there is *very high confidence* that the uncertainties in cloud processes explain much of the spread in modelled climate sensitivity" (AR5, Chapter 9, p. 3). It is even the only possible explanation, once the coefficients of radiative forcing and their uncertainties have been excluded from the models.

The width of the assumed range of the coefficient λ_G is clearly not negligible (see section 5.4), but it is obvious that the spread of IPCC simulations would have been incommensurably greater if it had reflected the uncertainties which truly apply to the entirety of physical knowledge, and not only to λ_G. Had it been a little narrower, we could have said, as a caricature, that the participants of the CMIP5 went to a lot of trouble for nothing: the global temperatures obtained were predetermined by the RCPs which they had to accept as input data for their simulations. The models can then describe the spatial distribution of the radiative balance with an unheard-of level of detail, as well as regional temperatures and the frequency of extreme climate events. At the output of the models can be found only what had been input. What is contested here is not the numerical realization of the general circulation models (nor the ability of their authors), but the validity of the fundamental input data made up of the RCPs, highly dependent on the assumed parameters of radiative forcing.

9.4. Comparative radiative forcing

Radiative forcing Q does not directly influence our energy balance models, but only in the form $Q = \lambda_G\ (S_1u_1 + S_2u_2 + S_3u_3)$, where the coefficient λ_G is unknown, and is not estimated. In the absence of radiative forcing Q, Figure 9.4 shows the sum $\Sigma = S_1u_1 + S_2u_2 + S_3u_3$ which is directly proportional to it (since in any case, the coefficient λ_G is of the same order of magnitude, with the unit, in °C/Wm^{-2}).

Here, we chose the combination of data selected for the partial results in Chapter 7. On the right, the sum Σ which results from forced identification with a limit on the sensitivity to solar activity closely resembles the "historic" RCPs. However, the graph on the left has nothing in common with this: here, the contribution of solar activity is predominant.

Due to the effects of filtering caused by the inertias of the climate system, forcings as different as that shown in Figure 9.5 nevertheless lead to the temperature reproductions observed, which are of comparable quality (see Figures 9.2 and 9.3), despite the natural/anthropogenic distribution being almost completely reversed.

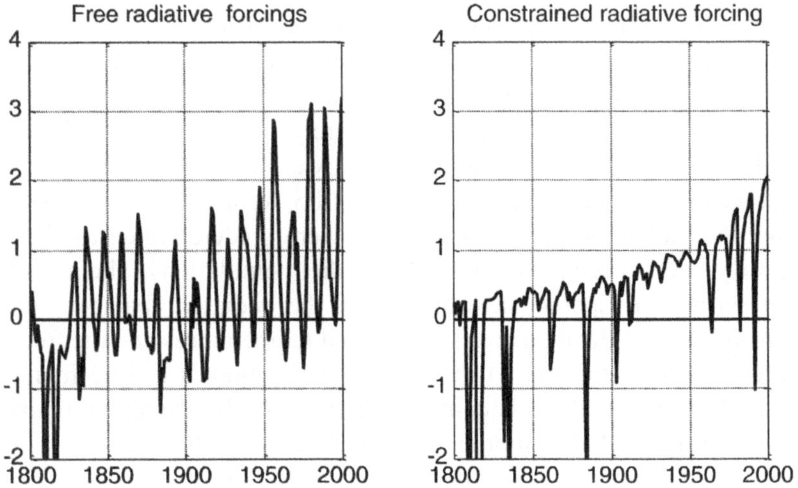

Figure 9.5. *For comparison with "historic" RCPs*

No conclusion could be reached if the millennium-old simulations were not available to make the difference. Figure 5.8 of the AR5 shows results of simulations where global temperature reconstructions are affected by fuzziness which is unfortunately hardly legible. As far as can be discerned, the simulated temperatures are closer to our Figure 7.5 regarding a forced identification than to Figure 7.2, which would doubtlessly reproduce large climatic events better, especially around the year 1000. Moreover, we will not go back over the statistical analysis which invalidate the IPCC's crucial hypothesis whereby sensitivity to solar activity is insignificant.

10

Long-term Climate Projections

10.1. IPCC scenarios and projections

With the RCPs, the IPCC advantageously replaced its previous emissions scenarios (SRES: Special Report on Emission Scenarios) with concentration scenarios. The difference is a radical one. It was legitimate to hypothesize on future human behavior, in particular CO_2 emissions, through the combustion of fossil fuels. It is less so to build a scenario on future CO_2 concentrations, since these do not solely depend on anthropogenic emissions, but on the whole of the "carbon cycle": the way in which nature absorbs, maintains or rejects the CO_2 according to the concentration which has been reached, and the present global temperature. In principle, two coupled models should be used: on one hand, the EBMs (or GCMs), where global temperature is an output and CO_2 concentration is one of the inputs. On the other hand, that of the carbon cycle whose output is CO_2 concentration and whose inputs are the emissions rate and global temperature.

An extremely simplistic carbon cycle model consists of assuming that a given fraction of anthropogenic emissions, known as the "airborne fraction" (AR5, section 6.3.2.4), is accumulated in the atmosphere, and that the rest is absorbed in various forms by the Earth's surface (solution in the oceans, biomass, carbonates and other "carbon sinks"). It is simplistic because the fractions absorbed and re-emitted obviously depend in part on the atmospheric concentration itself, as well as on the global temperature. When the IPCC announces that the so-called airborne fraction is 45%, this can only be true temporarily, during the fleeting concordance between the generalized phases of exponential growth which recently occurred (concurrently temperatures, emissions and CO_2 concentrations). Claiming that the transported fraction is variable and/or dependent on emissions is a systemic aberration, like that regarding λ_G in section 4.4. Nevertheless, it is tolerable to produce scenarios on future concentration, as long as it is clearly stated that the emission targets established to achieve them are highly dependent on the long-term validity of carbon cycle models, in particular temperature feedback on the absorption and re-emission of CO_2.

RCPs go even further than moving from emission scenarios to concentration scenarios. They directly impose scenarios in terms of radiative forcing.

Let us now return to the RCPs and ECPs (Extended Concentration Pathways) represented on the right of Figure 9.4. What we asserted (section 9.3) about "historic" radiative forcing remains true of the RCPs: they pre-incorporate all the coefficients of radiative forcing, without taking into account any uncertainties. Then the spread of simulated global temperature of the Figure 10.1 results only from the uncertainties of the factors of climate feedback in the simulated models. To the extent where this coefficient is nearly determined (each RCP predetermines the global temperatures involved) with scarcely any need to move on to the enormous simulations which use them.

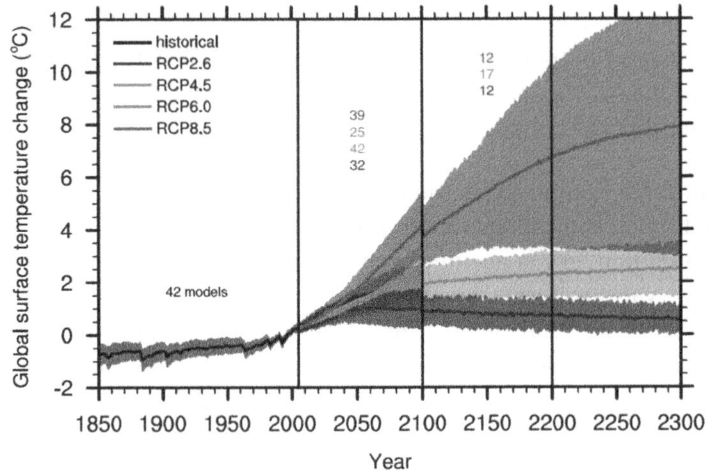

Figure 10.1. *IPCC long-term projections (AR5, Figure 12.5)*

Indeed, if we simply multiply the RCPs in Figure 9.4 by the assumed range of uncertainty ($0.27°C/Wm^{-2} < S_G < 1.62°C/Wm^{-2}$, see section 5.4) of the global sensitivity to radiative forcing S_G, we obtain – almost exactly – the projections in Figure 10.1. However, these detailed RCPs help to describe in the most accurate way the range of phenomena which would accompany the presupposed global climate evolution: rising oceans, melting ice caps, desertification, distribution of precipitation, and so on. The IPCC could talk endlessly on this point, but these simulations are only as valuable as the *a priori* knowledge entered into the RCPs, especially the assumed radiative forcing from the greenhouse effect and aerosols, and the lack of modeling of solar activity other than the direct energy factor. At the output of the models, we simply find what had been entered initially[1].

We shall now move on to "free" projections, where neither the models nor their input data are predetermined according to what message the IPCC want them to communicate.

10.2. EBM compatible scenarios

EBM simulations are only useful if we are free to carry out projections based on all of the parameters identified, including the radiative forcing coefficient α_i. To do so, it is fortunately possible to recover the corresponding concentrations of CO_2 available with the RCP provisions. For the scenario RCP 8.5, for instance, the final stabilized concentration of CO_2 reaches up to 7.4 times the pre-industrial concentration, hence the log-concentration line in Figure 10.2 (top graph, where: log(7.4) ~2).

To carry out projections, it is also necessary to anticipate solar activity. After the peaks of activity reached towards the end of the 20th Century, solar physicists expect a return to normal activity before the end of the 21st Century (if not a steep decline). Using standard signal processing techniques, it is easy to extrapolate a narrow-band signal. Therefore, we modeled solar irradiance via a Markov process, slightly damped around the 11 year period, and set with a return towards the millennium's average value (100 year time constant). A Kalman filter carries out a state estimation, and the

1 Garbage input, garbage output.

extrapolation results from the simulation of the free model from 2013 (Figure 10.2, middle graph).

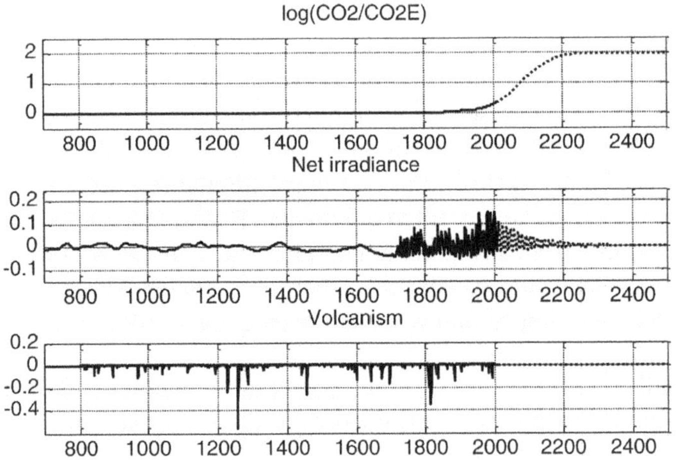

Figure 10.2. *Reconstructed (––) and anticipated (....) input signals*

Finally, since volcanic activity is unpredictable, an estimate is used based on its average value over the last millennium.

10.3. Long-term projections

In the previous chapters, in accordance with our option of leaving the identification free of all constraints, we did not want to assign a lower limit to anthropogenic sensitivity. Therefore, the estimated value is occasionally negative, which even the most radical climate skeptics dare not claim. In this chapter and the next, dedicated to future projections, the highest negative sensitivities identified would consequently lead to significant cooling, likely to give rise to pointless discussions. The free models will therefore include the weak constraint $S_{c\lim} > 0$. This is the only slight adjustment that we will make to the principle of identification free of all *a priori* knowledge, with the risk of being accused of lacking boldness in the exploration of what is possible.

Figure 10.3 shows two bundles of projections. The first (a) is obtained by simulating up to 2300 (like in Figure 10.1), with the input data in Figure 10.2

driving the freely identified models (with the weak constraint mentioned above). For forced simulations, we use the strong IPCC's constraint $S_{irr} < 1.62°C / Wm^{-2}$, leaving all the other parameters free and identified. The zero temperature in the figure is the conventional preindustrial equilibrium temperature (reference 1850–1900). We highlighted the future year 2100, as well as the critical level of a 2-degree warming, which is the basis of negotiations during international climate summits.

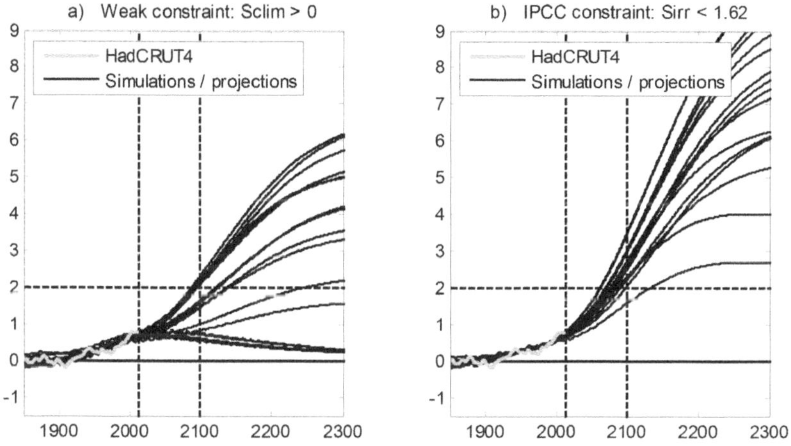

Figure 10.3. *Projections under RCP 8.5*

By severely limiting the contribution of solar activity, the IPCC harms the reproduction of millennia-old temperatures (see Chapter 7), and therefore indirectly reinforces the anthropogenic radiative forcing coefficient (see Chapter 9). Consequently, the bundle of curves in graph (b), resembles the one in Figure 10.1 (under RCP 8.5). In all cases but one, the projections to 2100 cross the line indicating a warming of two degrees above the conventional preindustrial temperature.

The range of free simulations (a) is much less alarming. In some cases, future temperatures decrease, when anthropogenic sensitivity is estimated as being zero, or that the drop in solar activity over-rides the effect of an insignificant anthropogenic sensitivity. In only a small number of cases, all of which come from the *Jones and Mann* reconstruction, is the critical threshold of 2°C exceeded – though only by a small margin – before the end of the century. Several centuries into the future, some projections may still become alarming, but their reliability depends on the accuracy of other factors, including the carbon cycle, and also on the absence of greenhouse effect saturation.

Should we use the term projections or extrapolations? In the case of CMIP5 the simulated models owe nothing to previous observations: future simulations owe almost everything to the scenarios selected and the pre-determined knowledge-based models. The term projection would therefore seem more appropriate.

In our case, it is legitimate to use the term extrapolation. The future remains dependent on the scenario of CO_2 concentration, but the model itself owes everything to the paleoclimatic observations. By means of comparison, when we carry out a linear extrapolation ($y(t+\tau)=a\tau+b$), coefficients a and b constitute, to some extent, a model, identified by linear regression of the recent past of the observed signal. In our case, the model (EBM) is more complex and features input from exogenous signals, but its coefficients are determined using the signal to be extrapolated, and are only involved as intermediary results in the calculation. The only *a priori* knowledge involved is the energy balance structure (no more arbitrary than the affine structure of the linear extrapolations), and the minor constraint $S_{\text{clim}}>0$, which is usually inactive.

The simulations in graph (b) also constitute extrapolations according to our reasoning, but they use the strict constraint $S_{irr}<1.62°C/Wm^{-2}$, which arises from an *a priori* certainty, without which the observations would contradict the alarmist assumptions of the IPCC.

The ranges in the two graphs in Figure 10.3 must not be interpreted as being representative of the calculated confidence intervals. They simply reflect the discordance of the various paleoclimatic reconstructions from which they are derived. The ranges indicated for the IPCC projections (Figure 10.1) are more formal (+/– 1.64 standard deviation). However, they are only as reliable as the prior uncertainties on the models' climatic reactions and, even earlier, on the radiative forcing incorporated in the RCPs (AR5, Figure 12.4).

10.4. A disaster scenario

Figure 10.4 shows a scenario of solar activity which could lead to a different type of climate disaster, prompting a return to the little ice age. In a small number cycles, solar activity reaches the level of the 1700s, known as the Maunder minimum.

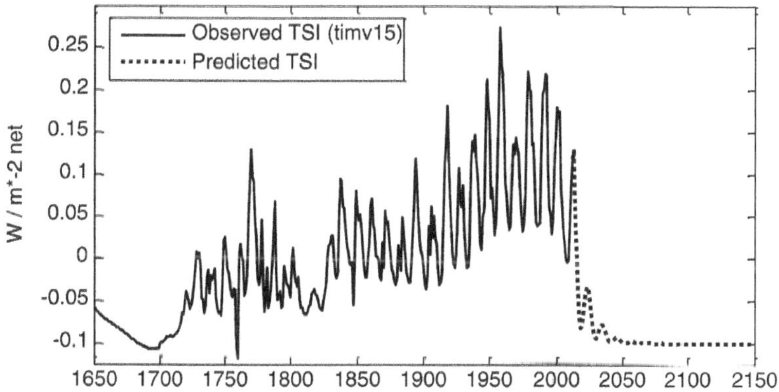

Figure 10.4. *Scenario: return to the Maunder level*

Graph (a) in Figure 10.5 shows simulations of the identified model under the constraint $S_{clin} > 0$. In some cases, the fall in solar activity initially has the greatest influence, then the rise in CO_2 is able to reverse the trend, thanks to a positive climate sensitivity.

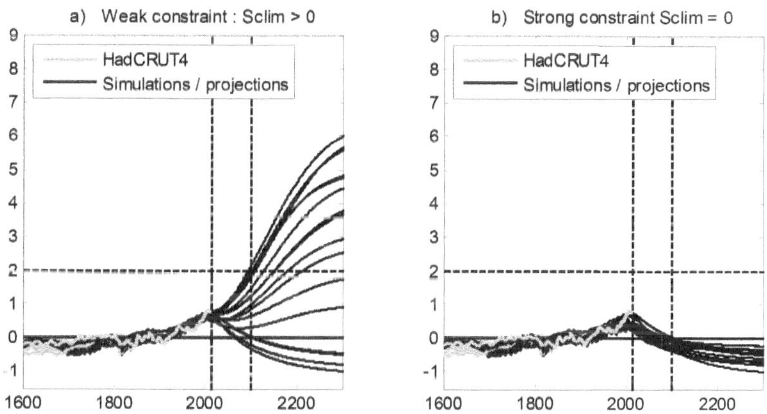

Figure 10.5. *Return to the little ice age*

The real climate disaster occurs in the event (admissible, see section 8.3) where climate sensitivity is zero or insignificant. If identification is forced by $S_{clim} = 0$, we can then obtain the simulation in graph (b), where the temperatures fall to the level of the little ice age (1600–1800), or even below it.

Let us therefore hope that solar activity does not remain at the level of the Maunder minimum for centuries. Or otherwise, that climate sensitivity is sufficient so that the gigatons of CO_2 released into the atmosphere save us from another LIA and the catastrophes it would bring with it.

11

Short-term Predictions

11.1. Decadal time scale predictions by GCM

Figure 11.1 below reproduces Figure 11.9 from the AR5. It gives a summary of the results from the CMIP5, and is based on the scenario RCP 4.5, which is less extreme than the RCP 8.5. Since we did not participate in the project, we are not aware of the detail of the protocols for the development of models, and the exact nature of the constraints to which the participants were subjected. The striking similarity between the simulations from the 42 models leads us to believe that these constraints are strict, starting with the obligatory use of the RCPs, which leaves no freedom in the evaluation of the radiative forcing coefficients – or the equivalent – in the GCMs.

The thin lines in graph (a), as well as the uniform shaded areas in (b) come from simulations initialized in 1986. From 1986 to 2005, they result from the so-called historic forcing, actually coming from coefficients of radiative forcing assumed by the IPCC, and after 2006, from the scenario RCP 4.5, based on the same assumed coefficients. The spread of the results is partly due to the nature of GCMs themselves, which, like numerical wind-tunnels, "naturally" simulate atmospheric or oceanic turbulence. These models are not explicitly based on a specific coefficient of climate reaction λ_G, but the weak spread amongst the simulated responses, to an imposed scenario of radiative forcing (RCP 4.5), suggests that the effective global feedback varies little from one model to the next.

The average of the simulations is indicated by the white median line in graph (b). The (perhaps overestimated) effect of radiative forcing caused by the eruption of Pinatubo in 1991 can be clearly distinguished. This effect aside, the simulated average global temperature is essentially due to anthropogenic forcing. The contribution of solar irradiance is imperceptible.

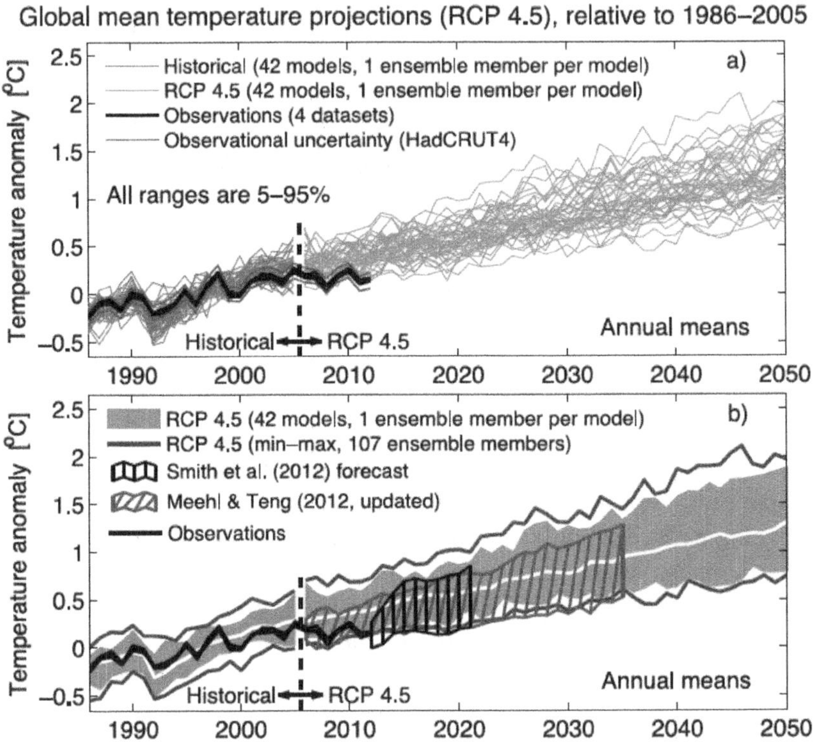

Figure 11.1. *Short-term projections from the IPCC*

The predictive power of these models is disappointing, if not non-existent: in 2012, the real temperature was below the overall envelope of all the simulations. It would clearly be wrong to criticize simulations started in 1986 for not reproducing the random fluctuations in later years, or the peak in 1998, due to an *El Nino* episode which could not have been predicted in 1986, or the climate "pause" which followed. However, although the latter is due to climatic fluctuations, the simulations should reproduce similar fluctuations here and there, because of the oceanic chaos generated by the simulations. This should have resulted in a bundle of projections with a spread such that the actual temperature does not leave this range, as it did in 2012.

Let us now study the bundle determined by Meehl and Teng (oblique hatching area), using an initialization updated in 2006. Although at the time the stagnation was hardly perceptible, it had already started, and the re-initialization could (or should) have provided a bundle of predictions able to remain accurate for at least 6 years, before being proven invalid by observations.

These failures are explained by the stance taken by the IPCC. It maintains that the dramatic rise seen up until 2000 is largely due to the anthropogenic factor, with the natural fluctuations deemed insignificant, and the volcanic factor even having some effect in the opposite direction. Curiously enough, on the medium white line in graph (b), there is no perceptible contribution from the Schwabe cycles. According to the correlation established by Camp (not to mention ours), solar activity could have contributed to some extent to the climate pause observed during the last decade. As weak – and likely – as it may be, such a contribution of solar activity to the climate pause would contradict the theory on the anthropogenic origin of the warming. This would be due not so much to its amplitude over the period 2000-2010, as to the fact that the high frequency/low frequency ratio of the models' dynamic (EMB or GCM) would imply an equilibrium solar sensitivity which could explain all or part of the warming observed over the 20th Century. It is understandable why the IPCC prefers to ignore the immediate role of the sun in the climate pause, however small it may be.

In this context, the IPCC can only attribute the current climate pause to a negative fluctuation in the climate's natural variability which should come to an end sooner or later. This clearly appears on the projection bundle by Smith *et al.* (vertical hatching area), which predicts, after 2013, a return to strong temperature increases.

11.2. The climate's natural variability

Climatic variations (and especially variations in global temperature) result either from duly identified causes, or from phenomena assumed to be negligible (the geothermal flow), or from causes omitted because of a lack of sufficient scientific data (geomagnetism), or simply those which are as yet unknown. They are also caused by anything with random attributes, intrinsic to the turbulent (chaotic) nature of tropospheric and oceanic flows, known as *internal variability*.

We accept that the random tropospheric fluctuations over a year cause average variations which are *a priori* independent from one year to the next. Oceanic chaos has an influence on an incomparably larger time scale. The large oceanic gyres of surface currents, with their turbulence, take years to complete their circuit. Thermohaline circulation is even slower. We know that this is caused by the downwellings of waters being cooled around the poles, until they rise again, far away and a long time after, in the middle of the ocean or at the coasts (the famous *El Nino / La Nina*). The molecules' journey in deep water in this case takes hundreds or thousands of years. The thermohaline circulation is nevertheless governed by intermediate time scales, either decadal or centennial. Indeed, oceanic time constants are not only linked to transport time: the erratic changes in polar descents cause local fluctuation in the thermocline[1] level, and the variations in this level are spread as much by waves as by the transport of matter, occasionally in the opposite direction to the current, like tidal propagation in estuaries (see the Saint-Venant equations). These deep waves, invisible on the surface, propagate and bounce back from one side of the ocean to the other, and upwellings result from these sorts of backwashes or surges, which are both slow and random throughout the decades.

On this issue, the indices of ENSO (El Nino Southern Oscillation), NAO (North Atlantic Oscillation), PNAO (Pacific North American Oscillation) are correlated with global surface temperature, and come before the temperature variations. They help to refine the very short-term climate predictions, but these are obviously effect rather than external causes of the climate. They are outputs and not inputs of the climate process overall.

The erratic fluctuations in the thermohaline circulation lead to a process of accumulation and restitution of heat between the deep oceanic layers and the surface, and thereby generate flows of energy which are added to the radiative forcing on top of the identified input factors. The identified EBM output errors result from internal variability, to which can be added:

– output measurement errors;

– the effects of errors of measurement and of input reconstruction, influencing the output via the models;

1 The thermocline is the zone of rapid thermal transition between surface and deep waters.

– the effects of model errors applied to inputs (structural and parametric errors).

A model having been established (*a priori* or identified) as well as the input scenarios, a long-term prediction is reduced to a simple simulation, while a short-term prediction requires an extrapolation of the output error to be added to the simulation, whether this is totally due to internal climate variability or not.

Consequently, let us revisit such an output error (Figure 11.2), extracted from Figure 7.6.

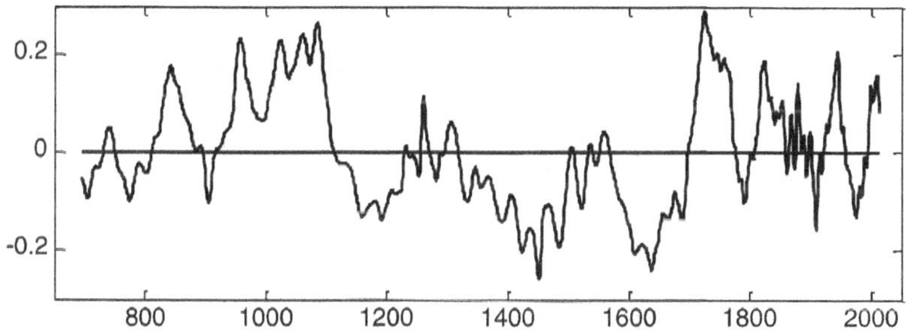

Figure 11.2. *The output error: image of random climate fluctuations*

The above line is zero mean, but it shows some continuity: it is not white noise. It shows no regular periodicity, apart from perhaps over the last two-hundred years, but such periodicity is dubious given that it was either absent or hidden for the previous 1000 years. It cannot be confirmed elsewhere on the correlation function (see Figure 7.10). At first glance, when the signal is clearly positive, we would expect it to return to the average value in 10 or 20 years. We are not able to go any further than this, and are left with a null prediction.

According to Figure 11.2, it appears that around the year 2000, the natural fluctuation was clearly positive. We could therefore expect a drop in this level in the coming years. The theory of the IPCC is precisely the opposite, maintaining that the current stagnation is due to the (temporary) cancelling out of a strong, deterministic warming effect (caused by CO_2) by a natural negative fluctuation. This point is contradicted, not only by Figure 11.2, but by all of the historic observations, processed together by all the models, free or forced. Indeed, let us consider Figure 11.3, which combines all of the 32 output errors: our 16 data combinations, firstly processed by free identifications, and then by identifications constrained by IPCC assumptions.

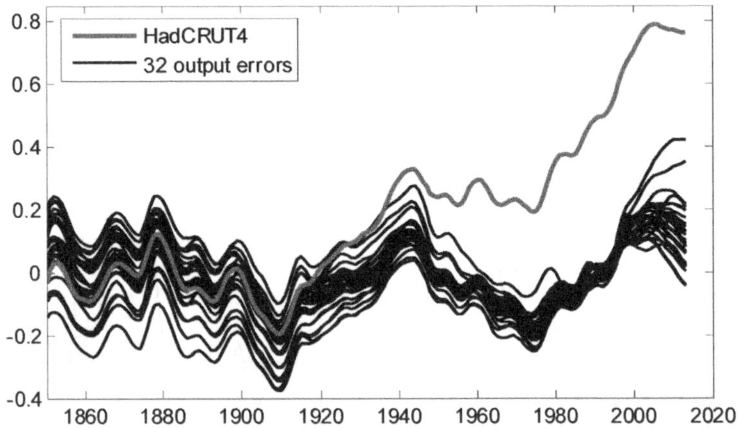

Figure 11.3. *Contribution of internal variability to global temperature*

The differences between the observed temperature (*HadCRUT*4) and output errors are constituted of the sum of the contributions determined by the identified inputs. The distribution of these contributions is very variable, depending on identification methods and combinations of paleoclimatic data. In any case, the residual output errors, images of natural climate variability, overlap each other remarkably well, especially over the last quarter of a century (1975-2000), which witnessed the majority of the "climate change". Therefore, it seems clear that the internal climate variability made a powerful contribution to warming, regardless of the other causes and their contributions: solar, volcanic or human activity.

Any attempt at a short-term prediction (over a ten-year period), which does not include an attempt to predict the internal variability is therefore clearly bound to fail.

11.3. State estimate and prediction

Any acceptable climate model (GCM or EBM) necessarily implies, in one way or another, the phenomena of accumulation and restitution of ocean heat. One of the essential components of the *state* of the climate process (state in the system theory sense) is made up of the quantity of heat accumulated by the oceans. If, sooner or later, this quantity rises – even very slightly – above its equilibrium level, the oceans would have to restitute it at some point. This would heat the surface, and the prediction of these phenomena can be made through an estimation of the state of the climate process.

We know that multi-day weather forecasts are based on knowledge of the atmospheric state observed from one day to the next, as much by terrestrial or maritime stations as satellites or weather balloons. In the context of short-term climate predictions (i.e. annual or decadal), it is also necessary to incorporate the state of the ocean, not only at the surface, but also at depth. Determining the initial state of GCMs is not a trivial matter, and the AR5 devotes a large table to the initialization methods of various GCM participants in the CMIP5. Even with the ocean observation system ARGO, we are still far from having a systematic initialization method, coherent with oceanic models which are still in their early stages.

Regarding our EBM structure, the rational transmittance $G(s)$ can be easily converted into state space equations. The form which is used is unimportant and may be modal, controllable/observable canonical or any other. Physically, each of the time constants of transmittance $G(s)$ corresponds to a thermal inertia: medium or deep ocean layers for the largest[2], atmosphere and ultra-superficial layer of oceans and continents for the smallest. In a "grey box" context, it would be tempting to determine some thermal inertia and exchange coefficients, in order to link – in both directions – with the identified coefficients. This hope is immediately dashed, given that we already had to give up hope of individually identifying coefficients α_i and λ_G, but only the relation $S_i = \alpha_i / \lambda_G$. Therefore, even if the components of the state vector correspond to quantities of heat accumulated, it is impossible to ascertain what these are with any more accuracy. As a next step, it may be possible to achieve this using *a priori* knowledge, but as long as they are not sufficiently reliable, we are obliged to stay with the "black box" approach.

2 Not forgetting the cryosphere.

Whatever the transfer-state conversion adopted for $G(s)$, we decided to estimate the state of the model by optimal statistical filtering (Kalman). To do so, as has already been done [HAS 76], we enhance the structure of the model with a noise of measurement v and random turbulence w, with respective power spectra r and q :

$$T_G = T_E + G(s)(Q + w) + v$$

We accept that noises of measurement v are independent from one year to the next, and we assess their standard deviation at around 0.05°C. Hence $r = 0.0025$.

The noise w added to the inputs Q is a catch-all. Firstly, it models the causes of internal variability. These are characterized by a zero power around frequency 0, due to the very principle of energy balance models, which cannot have long-term "missing heat". It also models measurement errors on forcing inputs u_i, where bias and calibration errors this time contribute to the zero frequency. Finally, for lack of anything better, we accept that the spectra of w is constant over a broad frequency band, determining its spectra q by using the following theoretical equality:

$$V_G = q\int_0^\infty g^2(t)\,dt + r$$

where V_G is the variance observed in the output error of the model and $g(t)$ the impulse response to the transmittance $G(s)$. In the end, the triplet $[G(s), r, q]$ contains all the data necessary for determining a state estimation filter. This will be statistically optimal only if v and w are indeed white noises of spectra r and q. In any case, whether optimal or not, it leads to a stable state observer, according to a rational procedure where no degree of freedom is left at the disposal of its designer: once again, the input-output data speaks for itself, without being oriented towards any presumption whatsoever.

The observer obtained in this way operates over the whole of the identification period. After which, from the prediction start date, the model continues to simulate, initialized by the estimated state and driven by anticipated input data.

All these technical operations were guided by a simple objective: linking the current observed temperature to long-term predictions.

11.4. Decadal time scale predictions by EBM

Figure 11.3 shows predictions which we made during 2014. The 2013 global temperature was already known for 2013, but not yet for 2014.

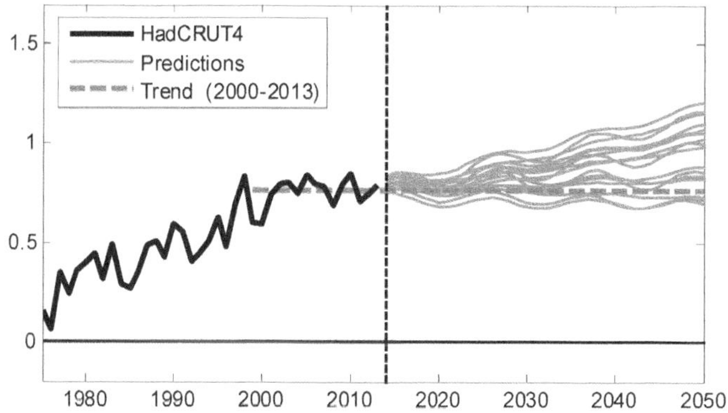

Figure 11.4. *Short-term predictions*

From these predictions, we can distinguish an oscillatory component resulting from the extrapolation of the effect of the Schwabe cycles. In the short-term (2020), we predict on average a small drop in temperature, which may be hidden by internal variability. In the medium-term, there appears to be no clear upwards trend like that of the decades 1970-2000. In the even longer term, the projections reach those in Figure 10.3, with their uncertainties, linked as much to the estimation of climate sensitivities as to predictions of solar activity.

The projections in Figure 11.4 are based on a collection of models whose coefficients of climate sensitivity are in the range [0, 2.48] °*C*, and the total sensitivity to irradiance in the range [1.9, 29.0] $°C/Wm^{-2}$. Since future temperatures will remain subject to natural and varied fluctuations, it will be necessary to wait for one or two decades more in order to start seeing the true figures. And even if one of the above models was "correct", there is nothing to guarantee that the causal inputs (solar activity and volcanic activity) will adhere to their predictions.

At this level, short-term predictions present nothing spectacular: one may even believe in a simple linear extrapolation using temperatures from the start of this century. This is not at all the case: if it were, the predictions which we will now carry out from the year 2000 should be just as inaccurate as those of the IPCC.

11.5. *A posteriori* predictions

Firstly, we will make a prediction for solar activity from the year 2000 (see section 10.2). Figure 11.5 compares the prediction with the irradiance actually observed until 2013.

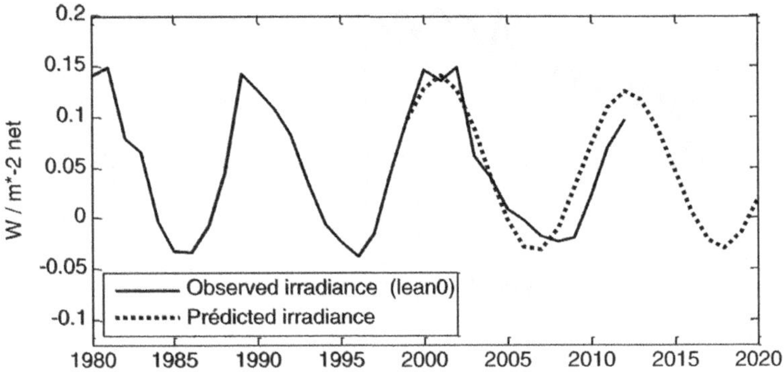

Figure 11.5. *Predicted and observed irradiance*

Of course, our prediction method (see section 10.2) does not pretend to compete with that of the solar physicists. It simply satisfies the requirement of making a plausible short-term prediction, which owes nothing to knowledge of a future which is as yet unknown.

For CO_2, we adopted the RCP 4.5 and, for volcanism, a prediction using the average value over the last millennium.

Figure 11.6 shows retroactive (or anticipated) predictions carried out from the year 2000, and these projections bear no resemblance to any kind of linear extrapolation. The observations do not leave the bottom of the bundle, like in Figure 11.1, where the anticipation does indeed start later (2006).

In 1999, we were not yet looking into the climate issue, but had that been the case, we would have been able to develop the identical identification algorithms and predictions shown here. We would therefore have obtained the same results[3], without knowing anything about the future evolution of the climate. Of course, no one would have given the slightest credence to predictions so far from the simple visual extrapolation of the warming observed until 1999.

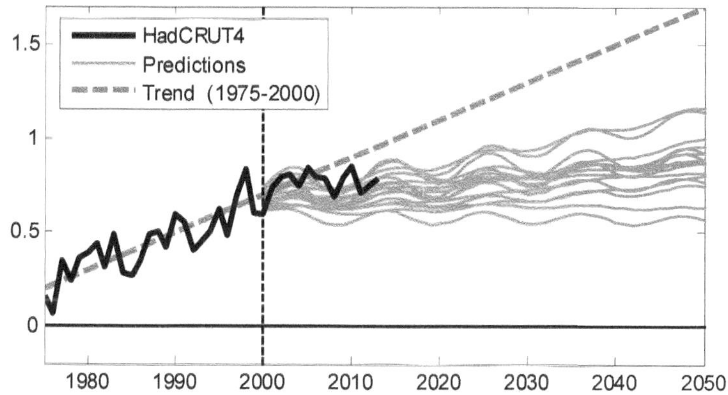

Figure 11.6. *Retroactive predictions*

We would highlight that, even with short-term predictions, they are based on models determined from data which covers more than a millennium. The deviations between the predictions for 2050 echo the deviations between the paleoclimatic data around the year 1000! In 2014, following the arrival of the climate pause, we were happily surprised to see the lines in Figure 11.6 appear as they do. One might have feared some disturbing effect of internal variability, and that the results are rather inconclusive. With *a posteriori* predictions in such close accordance with reality, we naively pondered how our own algorithm had done such a good job in reproducing these observations post-2000, without any prior knowledge. On reflection, the algorithm was able to express the combination of the following range of factors:

1) The turn-around in average solar activity had started long before (Figure 2.7), but its effect is only now being fully appreciated, due to climatic time constants;

2) Solar cycles reach minimum levels, during the decade 2002-2012;

3 Supposing paleoclimatic reconstructions were available at that time.

3) Internal climate variability reached its peak around 2000 (see Figure 11.3), and its subsequent short-term prediction can do nothing else but decrease;

4) In any case, human action identified is too weak to compensate – at least in the short-term – for the above downward trends.

Let us look again at point 3 above. By way of comparison, when a barometer is very high, we can predict – in the absence of any other information – that it will fall in the following days. The only error which could be made is on *when* this drop will take place. However, when the barometer gives an intermediate reading, it is not possible to predict whether the next movement will be up or down: the optimal prediction is the mean value (1014 mb), but this will be inevitably contradicted, in one direction or another, at some point in the future. *A posteriori*, it so happens that, by chance, the start date for the retroactive prediction (2000), corresponds to an extreme in internal variability. Another date (e.g. 1990), would have given less spectacular results but, in 2000, one would have to be willingly blinkered to predict anything other than a stagnation, if not a cooling at the start of the 21st Century.

Nevertheless, the IPCC had predicted nothing of the sort, and had no chance of predicting what actually happened at the start of the 21st Century, or even to explain it *a posteriori*, due to its refusal to take into account the nonetheless blatant solar activity. This is also due to its tendency to minimize the contribution of internal variability between the years 1980-2000, aiming to point the finger of blame squarely at CO_2. It has difficulty in explaining these circumstances. Indeed, this can be seen in the executive summary of Chapter 9 in the AR5, devoted to the evaluation of models (p.743):

"There is *very high confidence* that models reproduce the general features of the global-scale annual mean surface temperature increase over the historical period, including the more rapid warming in the second half of the 20th Century, and the cooling immediately following large volcanic eruptions[4]. Most simulations of the historical period do not reproduce the observed reduction in global mean surface warming trend over the last 10 to 15 years. There is *medium confidence* that the trend difference between the

4 Interestingly, the first part of this citation is in bold in the original text, but not the second.

models and observations between 1998-2012 is to a substantial degree caused by internal variability, with possible contributions from forcing error and some models overestimating the response to increasing greenhouse gas forcing."

It is strange to read a claim of "*very high confidence*" in what is in fact a simple observation: the reproduction by IPCC models of the rapid warming over the last half-century. The statement that the models fail to reproduce the current climate pause is also an observation, but the IPCC prefers to assign no level of confidence to this whatsoever. However, *medium confidence* is given to a diagnosis which arises from the evidence: internal variability, substantial forcing errors and overestimations of the greenhouse effect in the models, which could also apply to earlier periods.

Reading the entirety of Chapter 9 of the AR5, it seems that the IPCC is more and more circumspect with regard to its large models and their simulations – while maintaining their confidence in the most alarmist predictions[5]. We would highlight that this Chapter 9 concludes with a single, and rather revealing FAQ: "Are climate models getting better, and how would we know?" (p. 824). We can deliberate over the response:

"Climate models of today are, in principle, better than their predecessors. However, every bit of added complexity, while intended to improve some aspect of simulated climate, also introduces new sources of possible error (e.g. via uncertain parameters), and new interactions between model components that may, if only temporarily, degrade a model's simulation of other aspects of the climate system. Furthermore, despite the progress that has been made, scientific uncertainty regarding the details[6] of many processes remains."

5 Doublethink?

6 And the devil is in the detail.

12

Conclusions

12.1. On the identification

The very high confidence shown by the IPCC regarding the human origin of global warming is supported by an impressive theoretical corpus, and a collection of numerical simulations, carried out by dozens of research institutes, spending enormous computing resources.

Despite the established argument, this work intends to initiate a reliable methodology for macro-modeling of the climate process, using only historic observations; the objective being to determine a "black box-type" model based on global climate observations. While the modeling of industrial processes by identification is commonly carried out with success, identification of climatic processes presents difficulties and limits which are inherent to this approach[1]. The constraints on the issue practically impose the following fundamental options:

– a dynamic model with half a dozen parameters at most;

– continuous processing of climate archives over periods of the order of a millennium in duration;

– a method of identification by minimizing output error, accompanied by *ad hoc* statistical analysis;

1 This is may be the reason why we appear to be the first to fully apply the established theory of identification of dynamic systems, even though all the elements, models and climatic data would have allowed this to be done for some time.

– all of which is carried out without using any type of *a priori* knowledge.

We thus developed algorithms to allow the identification of a model incorporating equilibrium climate sensitivities with regard to the main potential causes given: CO_2 concentration, solar irradiance and volcanic activity. Estimates are accompanied by an assessment of associated uncertainties. The techniques proposed are easily accessible, and can be reproduced by anyone with some notion of systems theory. The programs which we developed (open source) are available to anyone wishing to use them[2].

We have obtained far-reaching results, especially regarding respective contributions of anthropogenic activity and solar activity to climatic variations. Identification also leads to models with remarkable powers of prediction.

12.2. Climate sensitivity

The major climate parameter is temperature sensitivity to a doubling in CO_2 levels. According to the consensus, the extreme limits range from 1°C to 6°C. Such a wide gap is inconceivable, after 25 years of IPCC work, as well as the self-satisfaction shown by heightened confidence levels announced in each five-yearly report.

Identification does not significantly reduce this range, but it moves it downwards: the confidence regions identified even spread to negative sensitivities. The uncertainty is due to the fact that there is a single and excessively short period (1975-2000), where CO_2 concentration and global temperature increased simultaneously. Indeed, we observed that over the past millennium, on many occasions, natural climate fluctuations often led to comparable temperature variations, both in terms of amplitude and duration (several tenths of a degree in a few decades). In such circumstances, not only is it impossible to accurately *estimate* sensitivity to CO_2, but we cannot even *detect* with any certainty the existence of a cause-effect relationship.

The assertion that human activity has a tangible impact results from theoretical concepts relating to the greenhouse effect, which it is perfectly

2 http://www.iste.co.uk/fr/delarminat.zip.

legitimate to call into question. We know that water constitutes the fundamental heat-transfer fluid in the climate machine, through its phase changes: oceans and clouds (liquid) ↔ gaseous (invisible). Greenhouse gases (including gaseous water) cannot act in isolation, without interacting with cloud formations (liquid water). However, the IPCC admits that the theoretical calculations regarding the greenhouse effect are valid in a supposedly clear sky (AR5, 8.3.1). In fact, there is no experimental confirmation of greenhouse effects by direct atmospheric measurements (weather balloon, satellites), and the AR5 is particularly discreet on the absence of a tropospheric hot spot which would nonetheless constitute its primary sign (AR4, Figure 9.1). In such conditions, the hypothesis that GHGs and human activity do not have a significant effect cannot be dismissed. We went on to a statistical test of this hypothesis, which shows that it cannot be rejected, with a significant certainty rate of 90%. We would not go as far as to say that this is the *most likely* situation, but, it is certain that anthropogenic sensitivity will remain very difficult to quantify for some time, both through theory and identification. Indeed, climate history being what it is, and given the forecasts of CO_2 emissions, we are unable to see how, only using observations, the accuracy of the sensitivity to CO_2 could improve in the foreseeable future. Paradoxically, this result is highly significant: it contradicts the assertion that climatic observations prove the anthropogenic origin of global warming.

12.3. Solar activity

The second fundamental climatic parameter is sensitivity to solar activity. In the context of identification, the situation is more positive: the past millennium saw successive large climatic variations: medieval warm period, little ice age and recent warming. There were also transitory changes, the most recent of which occurred around 1810, 1910 and 1940. Moreover, since 1850, the quality of the thermometric measurements helps us to discern the impact of 11-year solar cycles on temperatures. On all timescales, we see correlations between these climatic fluctuations and solar activity, as shown by measurements of irradiance, the observation of sun spots and archives of cosmogenic isotopes. Nevertheless, the natural variability of the climate clouds the perception of this: correlations which appear to be very revealing over some periods, are contradicted over others. Furthermore, the analysis of static correlation is misleading. As revealing as they may seem – precisely because they are static – the climatic cause-effect relationships are dynamic by their very nature. The theory of identification helps to directly handle this situation. It is not only effective, but indispensable, and also allows us to deal with the multi-causal nature of the climatic process other than with multiple static regressions. We will not reiterate the theoretical aspects of identification here, but as a reminder of the results: the analysis of uncertainty and hypothesis tests show that the contribution of solar activity to the evolution of global temperature is not a simple effect resulting from climatic variability. Instead, its level is, with high probability, up to ten times greater than the insignificant level assigned to it by the IPCC, on the basis of a single energy factor[3]. This probability is not just an "expert opinion", but is based on actual statistical calculations. It must be considered as a fact that solar activity, as a causal explanatory variable, does constitute the primary deterministic explanation of "climate change". Understanding the reasons for this action (cosmic activity or other) is another issue and does not fall within our remit.

12.4. Predictive capacity

In all case studies examined, the analysis of the simulations' output error shows that random climate fluctuations greatly contributed to the warming over the last half-century, whereas the IPCC would like to use natural variability to explain the current stagnation. The predictive power of the identified models confirms that the first hypothesis is correct one. Beyond accuracy calculations, nothing rivals the capacity of a model at predicting future behavior. Traditionally, for identification of industrial processes, a set of input-output data is kept for validation purposes. In this case, the experiment consists of carrying out identification on the database recorded until 2000 only, then comparing real temperatures which followed on with simulations. The test is conclusive: despite the uncertainty for identified parameters, the model reproduces the current temperature stagnation since the year 2000, even though nothing seemed to indicate such an outcome. In this regard, not one of the IPCC models made such a prediction, neither of the stagnation itself, nor of anything resembling it. In the possible scenario

3 Unless the solar irradiance reconstructions accepted by the IPCC are substantially reviewed, as Shapiro proposes.

of a return to weak solar activity as in the 1700s-1800s, our identified models instead give rise to fears of a drop in temperatures which could cancel out the rise of the past century. This event is neither more nor less probable than the alarming continuation of the rise predicted by the IPCC.

12.5. The climate change in question

Nobody denies that the 20th Century has seen temperatures increase by almost one degree, continuing a trend which started two centuries earlier. But is it possible to speak of anthropogenic, and more or less irreversible, climate change?

The analysis of the causes of this warming, as global identification highlights them, provides parts of the answer. A primary cause is solar activity. This is confirmed by the current stagnation, which belatedly results in a drop in solar activity which started 30 or 40 years ago. A second cause is the climate's natural variability: during the last quarter of a century, the ocean has released quantities of heat which have contributed to the warming, and the current stagnation can be explained by a recovery of this heat by the oceans. This recovery is probably far from being complete.

With regard to these two explanations, the effects of human activity are marginal, and difficult to assess via climatic observations. This could only be confirmed theoretically, but this seems even more uncertain, and there does not appear to be a real consensus on quantification.

In the future, a fall in solar activity could outweigh all other causes, taking away any significance from the expression *climate change* (and *a fortiori climate disruption*)

12.6. Prospects

Much still needs to be done to refine the quantification of identified models. Firstly upstream, for paleoclimatic reconstructions, where a consensus is far from being found, as well as for solar activity. Afterwards, work needs to be done on the identification techniques themselves. We went no further than the output error method, to ensure any manipulating by technique was impossible. Having established the predominant importance of solar activity, it would be logical to examine the evaluation of all climate parameters in a "grey box" context, introducing *a priori* knowledge, for instance regarding thermal inertia of ocean masses. Another possibility would be to continue the identification according to the approach initiated here, but based on a coupled climate-carbon model, with the exogenous input of the anthropogenic emission rate of CO_2, where the concentration becomes a state variable like temperature. Such work should be carried out by experts whose competence in the climate field is greater than ours.

Bibliography

[ADE 64] ADEM J., "On the physical basis for the numerical prediction of monthly and seasonal temperatures in the troposphere-ocean-continent system", *Monthly Weather Review,* vol. 92, pp. 91–103, 1964.

[ALM 13] ALMEIDA J., *et al.* "Molecular understanding of sulphuric acid-amine particle nucleation in the atmosphere", *Nature*, vol. 502, pp. 359–363, 2013.

[ARR 96] ARRHENIUS S., "XXXI. On the influence of carbonic acid in the air upon the temperature of the ground", *The London, Edinburgh, and Dublin Philosophical Magazine and Journal of Science,* vol. 41, pp. 237–276, 1896.

[BAR 00] BARD E.G., RAISBECK F., YIOU F., *et al.*, "Solar irradiance during the last 1200 years based on cosmogenic nuclides", *TELLUS B,* vol. 52, no. 3, pp. 985–992, 2000.

[BAR 03] BARD E. *et al.*, "Reconstructed solar irradiance data", IGBP PAGES/World Data Center for Paleoclimatology Data Contribution Series #2003-006, NOAA/NGDC Paleoclimatology Program, Boulder CO, 2003.

[BAR 97] BARD E., *et al.* "Solar modulation of cosmogenic nuclide production over the last millennium: comparison between ^{14}C and ^{10}Be records", *Earth and Planetary Science Letters,* vol. 150, pp. 453–462, 1997.

[BAR 07] BARD E., *et al.* "Comment on 'solar activity during the last 1000yr inferred from radionuclide records' by Muscheler *et al.* (2007)", *Quaternary Science Reviews,* vol. 26, pp. 2301–2304, 2007.

[BOE 05] BOER G.J., Hamilton K., Zhu W., "Climate sensitivity and climate change under strong forcing", *Climate Dynamics,* vol. 24, pp. 685–700, 2005.

[BUD 69] BUDYKO M.I., "The effect of solar radiation variations on the climate of the earth", *Tellus,* vol. 21, pp. 611–619, 1969.

[CAM 07] CAMP C.D., KA KIT T., "Surface warming by the solar cycle as revealed by the composite mean difference projection", *Geophysical Research Letters*, vol. 34, 2007.

[CHR 12] CHRISTIANSEN B., LJUNGQVIST F.C., "The extra-tropical northern hemisphere temperature in the last two millennia: reconstructions of low-frequency variability", *Climate of the Past*, vol. 8, 2012.

[CRO 00] CROWLEY T.J., "Causes of climate change over the past 1000 years", *Science*, vol. 289, pp. 270–277, 2000.

[CRO 13] CROWLEY T.J., MATTHEW B.U., "Technical details concerning development of a 1200 yr proxy index for global volcanism", *Earth System Science Data*, vol. 5, pp. 187–197, 2013.

[DEL 09] DE LARMINAT P., "LTI systems and transmittances", *Automatique Appliquée Hermès*, Second edition, 2009.

[DEL 11] DELAYGUE G., EDOUARD B., "An Antarctic view of beryllium-10 and solar activity for the past millennium", *Climate Dynamics*, vol. 36, pp. 2201–2218, 2011.

[ETH 98] ETHERIDGE D.M., *et al.*, Historical CO2 record derived from a spline fit (75 year cutoff) of the Law Dome DSS, DE08, and DE08-2 ice cores. Division of Atmospheric Research, 1998.

[FRI 91] FRIIS-CHRISTENSEN E., KNUD L., "Length of the solar cycle – an indicator of solar activity closely associated with climate", *Science*, vol. 254, pp. 698–700, 1991.

[FRO 04] FRÖHLICH, C., JUDITH L., "Solar radiative output and its variability: evidence and mechanisms", *The Astronomy and Astrophysics Review*, vol. 12, pp. 273–320, 2004.

[GAO 08] GAO C., ALAN R., CASPAR A., "Volcanic forcing of climate over the past 1500 years: an improved ice core-based index for climate models", *Journal of Geophysical Research: Atmospheres (1984–2012)*, vol. 113, no. D23, 2008.

[GHI 76] GHIL M., "Climate stability for a sellers-type model", *Journal of the Atmospheric Sciences*, vol. 33, pp. 3–20, 1976.

[HAN 85] HANSEN J., *et al.*, "Climate response times: dependence on climate sensitivity and ocean mixing", *Science*, vol. 229, pp. 857–859, 1985.

[HAN 02] HANSEN J., *et al.*, "Climate forcings in Goddard Institute for space studies SI2000 simulations", *Journal of Geophysical Research: Atmospheres (1984–2012)*, vol. 107, no. D18, 2002.

[HAN 05] HANSEN J., *et al.*, "Efficacy of climate forcings", *Journal of Geophysical Research: Atmospheres (1984–2012)*, vol. 110, no. D18, 2005.

[HAS 76] HASSELMANN K., "Stochastic climate models part I. Theory", *Tellus*, vol. 28, pp. 473–485, 1976.

[JÄV 10] JÄRVINEN H., *et al.*, "Estimation of ECHAM5 climate model closure parameters with adaptive MCMC", *Atmospheric Chemistry and Physics*, vol. 10, pp. 9993–10002, 2010.

[JON 13] JONES G.S., PETER A., STOTT, *et al.*, "Attribution of observed historical near–surface temperature variations to anthropogenic and natural causes using CMIP5 simulations", *Journal of Geophysical Research: Atmospheres*, vol. 118, pp. 4001–4024, 2013.

[JON 04] JONES P.D., MICHAEL E.M., "Climate over past millennia", *Reviews of Geophysics*, vol. 42, 2004.

[KIR 11] KIRKBY J., *et al.*, "Role of sulphuric acid, ammonia and galactic cosmic rays in atmospheric aerosol nucleation", *Nature*, vol. 476, pp. 429–433, 2011.

[KOP 05] KOPP G., GEORGE L., GARY R., "The total irradiance monitor (TIM): science results", *The Solar Radiation and Climate Experiment (SORCE)*, Springer, New York, pp. 129–139, 2005.

[KOS 13] KOSOLAPOVA E., *et al.*, "Summary of the 12th session of working group 1 of the intergovernmental panel on climate change (IPCC) and thirty-sixth session of the IPCC", *Earth Negotiations Bulletin*, vol. 12. no. 581, 29 September 2013.

[LEA 00] LEAN J., "Evolution of the sun's spectral irradiance since the Maunder minimum", *Geophysical Research Letters*, vol. 27, pp. 2425–2428, 2000.

[LEA 04] LEAN J., Solar irradiance reconstruction. IGBP PAGES/World Data Center for Paleoclimatology, Data Contribution Series# 2004-035, NOAA/NGDC Paleoclimatology Program, Boulder CO, 2004,

[LE 07] LE R.E., *Abrégé d'histoire du climat: du Moyen Âge à nos jours*, Fayard, 2007.

[LE 04] LE T.H., *Science du changement climatique. Acquis et controversies*, p. 32, 2004.

[LIN 01] LINDZEN R.S., MING-DAH C., ARTHUR Y.H., "Does the earth have an adaptive infrared iris?", *Bulletin of the American Meteorological Society*, vol. 82, pp. 417–432, 2001.

[LOE 07] LOEHLE C., "A 2000-year global temperature reconstruction based on non-tree ring proxies", *Energy & Environment*, vol. 18, pp. 1049–1058, 2007.

[LJU 99] LJUNG L., *System Identification: Theory for the User*, PTR Prentice Hall Information and System Sciences Series, Prentice Hall, 1999.

[LJU 10] LJUNGQVIST F.C., "A new reconstruction of temperature variability in the extra-tropical northern hemisphere during the last two millennia", *GeografiskaAnnaler: Series A, Physical Geography*, vol. 92, pp. 339–351, 2010.

[MAN 98] MANN M.E., Bradley R.S., Malcolm K., "Hughes, Global-scale temperature patterns and climate forcing over the past six centuries", *Nature*, vol. 392, pp. 779–787, 1998.

[MAN 99] MANN M.E., RAYMOND S.B., MALCOLM K.H., "Northern hemisphere temperatures during the past millennium: inferences, uncertainties, and limitations", *Geophysical Research Letters*, vol. 26, pp. 759–762, 1999.

[MAS 13] MASSON D.V., SCHULZ M., ABE-OUCHI A., *et al.*, "Information from Paleoclimate Archives", in STOCKER T.F., QIN D., PLATTNER G.-K., *et al.* (eds.), *Climate Change 2013: The Physical Science Basis. Contribution of Working Group I to the Fifth Assessment Report of the Intergovernmental Panel on Climate Change,* Cambridge University Press, Cambridge, UK and New York, NY, 2013.

[MAS 10] MASTRANDREA M.D., *et al.*, "Guidance note for lead authors of the IPCC fifth assessment report on consistent treatment of uncertainties", *Intergovernmental Panel on Climate Change (IPCC)*, 2010.

[MEI 11] MEINSHAUSEN M., RAPER S.C.B., WIGLEY T.M.L., "Emulating coupled atmosphere-ocean and carbon cycle models with a simpler model, MAGICC6–Part 1: model description and calibration", *Atmospheric Chemistry and Physics,* vol. 11, pp. 1417–1456, 2011.

[MEI 11] MEINSHAUSEN M., *et al.*, "The RCP greenhouse gas concentrations and their extensions from 1765 to 2300", *Climatic Change,* vol. 109, pp. 213–241, 2011.

[MIS 07] MISKOLCZI F.M., "Greenhouse effect in semi-transparent planetary atmospheres", *Quarterly Journal of the Hungarian Meteorological Service,* vol. 111, pp. 1–40, 2007.

[MOB 05] MOBERG A., *et al.*, "Highly variable northern hemisphere temperatures reconstructed from low-and high-resolution proxy data", *Nature,* vol. 433, pp. 613–617, 2005.

[MOB 08] MOBERG A., REZWAN M., THORSTEN M., "Analysis of the Moberg *et al.* (2005) hemispheric temperature reconstruction", *Climate Dynamics,* vol. 31, pp. 957–971, 2008.

[MOR 12] MORICE C.P., *et al.*, "Quantifying uncertainties in global and regional temperature change using an ensemble of observational estimates: the HadCRUT4 data set", *Journal of Geophysical Research: Atmospheres (1984–2012)*, vol. 117, 2012.

[MYH 98] MYHRE G., *et al.*, "New estimates of radiative forcing due to well mixed greenhouse gases", *Geophysical Research Letters,* vol. 25, pp. 2715–2718, 1998.

[NOR 75] NORTH G.R., "Theory of energy-balance climate models", *Journal of the Atmospheric Sciences,* vol. 32, pp. 2033–2043, 1975.

[RAM 97] RAMASWAMY V., CHEN C.-T., "Linear additivity of climate response for combined albedo and greenhouse perturbations", *Geophysical Research Letters,* vol. 24, pp. 567–570, 1997.

[RIT 10] RITTAUD B., *Le Mythe climatique*, Le Seuil, 2010.

[SAT 93] SATO M., *et al.*, "Stratospheric aerosol optical depths, 1850–1990", *Journal of Geophysical Research: Atmospheres (1984–2012),* vol. 98, pp. 22987–22994, 1993.

[SCA 14] SCAFETTA N., "The complex planetary synchronization structure of the solar system", *Pattern Recognition in Physics,* vol. 2, pp. 1–19, 2014.

[SEL 69] SELLERS W.D., "A global climatic model based on the energy balance of the earth-atmosphere system", *Journal of Applied Meteorology*, vol. 8, pp. 392–400, 1969.

[SHA 11] SHAPIRO A.I., SCHMUTZ W., ROZANOV E., *et al.*, "A new approach to long-term reconstruction of the solar irradiance leads to large historical solar forcing", *Astronomy & Astrophysics*, vol. 529, pp. 1–8, 2011.

[STA 11] STAUNING P., "Solar activity-climate relations: a different approach", *J. Atm. Solar-Terr. Phys.,* vol. 73, pp. 1999–2012, 2011.

[STE 09] STEINHILBER F., BEER J., FRÖHLICH C., "Total solar irradiance during the Holocene", *Geophysical Research Letters,* vol. 36, 2009.

[STO 13] STOCKER T.F., QIN D., PLATTNER G.-K., *et al.* (eds.), *IPCC: Climate Change 2013: The Physical Science Basis. Contribution of Working Group I to the Fifth Assessment Report of the Intergovernmental Panel on Climate Change*, Cambridge University Press, Cambridge, UK and New York, NY, 2013.

[STO 06] STOWASSER M., KEVIN H., GEORGE J., *et al.*, "Local and global climate feedbacks in models with differing climate sensitivities", *Journal of Climate*, vol. 19, 2006.

[SVE 97] SVENSMARK H., EIGILFRIIS C., "Variation of cosmic ray flux and global cloud coverage–a missing link in solar-climate relationships", *Journal of Atmospheric and Solar-Terrestrial Physics,* vol. 59, pp. 1225–1232, 1997.

[SVE 00] SVENSMARK H., "Cosmic rays and Earth's climate", *Space Science Reviews,* vol. 93, pp. 175–185, 2000.

[TAY 12] TAYLOR K.E., RONALD J.S., GERALD A.M., "An overview of CMIP5 and the experiment design", *Bulletin of the American Meteorological Society,* vol. 93, 2012.

[TRE 09] TRENBERTH K.E., FASULLO J., "Earth's global energy budget", *Bulletin of the American Meteorological Society*, vol. 90, pp. 311–323, 2009.

[USO 08] USOSKIN I.G., MONIKA K., KOVALTSOV G.A., "Role of centennial geomagnetic changes in local atmospheric ionization", *Geophysical Research Letters,* vol. 35, 2008.

[VAN 13] VAN HATEREN J.H., "A fractal climate response function can simulate global average temperature trends of the modern era and the past millennium", *Climate Dynamics,* vol. 40, pp. 2651–2670, 2013.

[VAN 11] VAN VUUREN D.P.J., EDMONDS A., KAINUMA M., *et al.*, "A special issue on the RCPs", *Climatic Change*, vol. 109, no. 1–2, pp. 1–4, November 2011.

[WEL 14] WELLER S.R., SCHULZ B.P., NINNESS B.M., "Identification of linear climate models from the CMIP3 multimodel ensemble", *19th IFAC World Congress,* Cape Town, South Africa, 24–29 August 2014.

[WIG 05] WIGLEY T.M.L., *et al.*, "Effect of climate sensitivity on the response to volcanic forcing", *Journal of Geophysical Research: Atmospheres (1984–2012),* vol. 110, 2005.

[WIL 03] WILLSON R.C., MORDINOV A.V., "Secular total solar irradiance trend during solar cycles 21 and 22", *Geophys. Res. Let.*, vol. 30, pp. 1199–1202, 2003.

Index

M, N

O, P, R

S, T, V

Printed and bound by CPI Group (UK) Ltd, Croydon, CR0 4YY

13/09/2026

14961016-0006